DANCE OF THE PHOTONS

DANCE OF THE PHOTONS

FROM EINSTEIN TO QUANTUM TELEPORTATION

ANTON ZEILINGER

FARRAR, STRAUS AND GIROUX

NEW YORK

Farrar, Straus and Giroux
18 West 18th Street, New York 10011

Distributed in Canada by D&M Publishers, Inc.
Printed in the United States of America
First edition, 2010

Originally published in 2005 in different form in German by C. Bertelsmann,
Germany, as *Einsteins Spuk: Teleportation und weitere Mysterien der Quantenphysik*.

Library of Congress Cataloging-in-Publication Data
Zeilinger, Anton.
  [Einsteins Spuk. English]
  Dance of the photons : from Einstein to quantum teleportation / Anton
Zeilinger.— 1st ed.
    p.   cm.
  Includes index.
  ISBN 978-0-374-23966-4 (hardcover)
  1. Quantum theory.   2. Quantum teleportation.   I. Title.

QC174.12 .Z4513 2010
530.12—dc22

                                                    2009031890

Designed by Jonathan D. Lippincott

www.fsgbooks.com

10   9   8   7   6   5   4   3   2   1

CONTENTS

DANCE OF THE PHOTONS

PROLOGUE: UNDERNEATH THE DANUBE

Every January 1, the New Year's Concert of the Vienna Philharmonic ushers in a new year. This concert is held in the great Golden Hall of the Musikverein, the home of the traditional music society of Vienna, and is transmitted worldwide to literally hundreds of millions of people eager to listen to the beautiful waltzes, polkas, overtures, and other pieces by the Strauss family and their contemporaries. Once the official program ends, we join the audience in applauding, but everyone is still waiting for the encore. Then the very low sounds of the strings start, and everyone applauds again, recognizing the expected piece. The orchestra stops, and the conductor wishes everyone in the concert hall and around the world a happy new year. Again, the strings start and the orchestra plays what is often called the unofficial Austrian anthem, the famous "Blue Danube" waltz by Johann Strauss the Younger. There are not many pieces of music that are able to convey both the pleasure and the intrinsic melancholy of human existence as well as this music, written for the grand balls in Vienna's imperial and courtly ballrooms and still performed today during the ball season every year.

Little do those present and those at their TV sets know that not far from the Golden Hall, within the city limits of Vienna, an experiment is being conducted at the cutting edge of modern technology, challenging the imagination with ideas previously found only in science fiction and with the implications of those ideas for how we can understand the world around us.

The concert ends with its final encore, Johann Strauss the Elder's "Radetzky March," one of the most vibrant and jolly pieces ever written. We leave the concert hall and drive to the river Danube. It is a beautiful winter day with not many people about, as January 1 is a national holiday. The Danube passes through the city of Vienna in two branches, forming a long island in between. We cross from one of the riverbanks onto the island over a bridge that not even our car's GPS knows, as it is not open to the public. The island is off limits to cars except for those on official business.

On the island, we head for a building hidden behind high trees. This is the location of the pumping station of Vienna's sewage system. There is a huge sewer passing under the river, connecting the two sides. Its purpose is to convey all the sewage collected on the eastern side of the river, a part of the city the Viennese lovingly call Transdanubien ("the place across the Danube"), to a huge waste-treatment plant on the other side. In this way, the Viennese, who are very environmentally conscious, make sure that sewage is not deposited directly into the Danube.

We enter the building and take the elevator two levels down, below the river. After a short walk, we reach two large tunnels opening to the left and the right, connecting both banks of the river, Transdanubien and Vienna proper. Through this huge tunnel, tubes run in parallel, carrying sewage, and there are many cables. Tucked away, near the entrance to one of the tunnels, a different scene greets us.

We see a small room off in a corner, with glass walls. Coming closer, we see laser light inside, with lots of high-tech equipment including modern electronics, computers, and the like, and we meet Rupert. He tells us that he is a student at the University of Vienna working on his Ph.D. dissertation, which he hopes to finish soon in order to earn his doctorate. The title of his dissertation is "Long-Distance Quantum Teleportation." We ask Rupert to briefly explain what it is that we see here. He tells us that the point of the experiment is to teleport a particle of light—a photon—from the Danube Island side of the riverbank over to the Vienna side.

Noticing that we don't understand much, he tells us that teleportation is a little bit like "beaming" in science fiction, "but not quite." He smiles broadly and starts to explain. While we still don't understand much, we listen with increasing fascination. He promises to give us a more detailed explanation later. At the moment, we just want to gain a small degree of familiarity with the language used, to get accustomed to the setup and the general concepts being studied, and to acquaint ourselves with the strange surroundings.

The lasers, we learn, are mainly here to produce a very special kind of light. Light consists of particles called photons, and this laser produces peculiar pairs of photons that are "entangled" with each other. This entanglement, as we shall learn in more detail later on, means that the two photons are intimately connected with each other. When one is measured, the state of the other one is instantly influenced, no matter how far apart they are separated.

The notion of *entanglement* was identified by the Austrian physicist Erwin Schrödinger in 1935. He wanted to characterize a very interesting state of affairs. Shortly before, Albert Einstein had hinted at an interesting new situation emerging in quantum mechanics in a paper he published together with his young colleagues Boris Podolsky and Nathan Rosen.

For us to understand a little of what entanglement is about, let's consider two particles that have had some interaction with each other. For example, they could have hit each other just as billiard balls do and could now be moving apart. In classical — that is, traditional — physics, if one billiard ball moves, say, to the right, the other one moves to the left. Furthermore, if we know the speed of the hitting ball and how it hit the ball at rest, and if we also know how fast and in which direction the ball that was at rest moves away, we can figure out exactly where the other ball goes. This is what a good billiard player actually does when he is figuring out how to hit a ball with his cue.

Quantum "billiard balls" are much stranger. They will also move away from each other after the collision, but with these interesting and very strange differences. Neither of the two balls has a well-defined speed, nor does it move in a specific direction. Actually, neither of the balls has a speed or direction after the collision. They just move apart from each other.

The crucial point is this: as soon as we observe one of the quantum billiard balls, the ball instantly assumes a certain speed and moves along a certain direction away from the collision. At that very moment — but not before — the other ball assumes the corresponding speed and direction. And this happens no matter how far apart the two balls are.

So, quantum billiard balls are entangled. Of course, this kind of phenomenon has not been seen for real billiard balls yet, but for elementary particles, it is standard fare. Two particles that collide with each other are still intimately connected over a large distance. The actual act of *observation* of one of the two particles influences the other one instantly, no matter how far away the other one is.

Einstein did not like this strange feature, and he called it "spooky action at a distance." He was hoping that physicists might find a way to get rid of the spookiness. In contrast to Einstein, Schrödinger accepted this feature as something completely new, and he coined the term "entanglement" for it. Entanglement is *the* feature of the quantum world that forces us to say farewell to all our cherished views of how the world is built up.

When we ask Rupert about the purpose of his entangled photons, he smiles and tells us, "That's the magic trick." He keeps one of the two photons at his mini-laboratory down below the level of the water and sends the other photon along a glass fiber to the receiver at the other side of the river.

Rupert talks about "Alice" and "Bob" sending photons to each other and talking to each other as if they were humans. But it turns out that they are imaginary experimentalists, Alice sitting in her laboratory here and Bob on the other side of the river.

When we ask Rupert why he calls these two Alice and Bob, he tells us that this is not his invention. The names come from the cryptography community, in which it is important to make sure that messages sent between two people cannot be read or heard by unauthorized third parties. We immediately think of spies in an exciting setting, but Rupert calms us down. Cryptography, he explains, is broadly used these days. Even if you log on to the Internet and transmit, say, your credit card number, it is usually encrypted so nobody else can read it. He continues: "Initially, people called the sender of the message 'A' and the receiver 'B,' and then someone thought it better to simply call them 'Alice' and 'Bob,' to make it easier to talk about them."

Rupert shows us the thin glass fiber where Bob's photon enters, apparently no different from those widely used in telecommunication these days.

We let our eyes follow the glass fiber cable from Rupert's laser through the wall of his small laboratory up to a place where it joins all the other cables running through the large tunnels under the Danube. Rupert follows our eyes and asks, "Want to see where it goes?" We eagerly say yes, and our small excursion to the underground of Vienna begins.

First, we enter a tube of about four meters (thirteen feet) in diameter that goes steeply downward. Below us are two pipes, each about a meter in diameter, which carry the sewage. As they are tightly sealed, this does not influence our comfort very much, though a little bit of a strange smell hangs in the air. We are easily able to walk upright, but the space is not very wide. To our right and left are cable trays. Somewhere on one of these cable trays is our small optical fiber. One of us remarks, "Just like *The Third Man*," reminded us of one of the greatest movies of all time, set in Vienna after World War II. Some of the movie's best scenes are wild chases in the city's underground sewage system. We expect Orson Welles to pop around the corner at any moment, and the Harry Lime theme played by Anton Karas on his zither seems to ring in our ears.

After some time, we reach the deepest point of our travels, and Rupert tells us that the river is just above us. It is difficult to avoid imagining what would happen if somehow a crack were to appear and the water of the river started to flood in. Which direction would we run in? Fortunately, nothing happens, and we continue trotting along. The path starts to climb slightly upward. After a while, we emerge into a small room, and looking out, we see we have passed under not only the river but also a little adjacent park, a railroad, and a major road.

In the room, the glass fiber leaves its plastic housing and ends up in a setup similar to the one on the island, but much smaller. Again, a computer is nearby, as are a few optical elements such as mirrors and prisms and lots of electronics. Rupert explains that what happens here is the measurement of the teleported photon and in particular the verification of whether it has all its properties and features intact. Of the cables leading to Rupert's small table, we see one running upward; it ends on the roof of the building we are in. Rupert proudly tells us that this is the "classical" channel connecting Alice and Bob—a standard radio connection between the two players. At this point, we are slightly confused. What is this classical channel for? What was Rupert talking about when he mentioned entangled photons? What is teleportation?

Before exploring these questions, we climb to the roof of the building and are rewarded with a great view. On the other side of the river is the building where Alice is located. The river flows rather swiftly in between. Ships pass by, making their own slow and steady progress. A few ducks and swans enjoy the clean water. On our side of the river, next to the building where we are, we see a little pagoda built by Vienna's Buddhist community, and immediately our minds drift off into philosophical questions like what might all this mean, what is our role in the universe, what are we doing when we observe the world, and what the heck does quantum physics have to do with all of this?

To the west, we see the hills of the Vienna Woods, which are actually the easternmost reaches of the Alps, and to the east, the edge of the great Hungarian plains. History drifts into our thoughts; we remember the fact that the Turks, coming from the east, twice tried unsuccessfully to conquer Vienna. We can imagine how a successful conquering of Vienna would have changed history. We also consider how the kinds of questions we ask, the very deep questions, those about the meaning of our existence, might depend on our culture—Buddhist, Islamic, Christian. It is getting cold, and we allow ourselves to return slowly back to the life of modern Vienna.

SPACE TRAVEL

When we hear of teleportation, we often think it would be an ideal means of traveling. We would simply disappear from wherever we happened to be and reappear immediately at our destination. The tantalizing part is that this would be the fastest possible way of traveling. Yet, a warning might be in order here: teleportation as a means of travel is still science fiction rather than science.

Thus far, people have only been able to travel to the Moon, which on a cosmic scale is extremely close, the equivalent of our backyard. Within our solar system, the closest planets, Venus and Mars, are already roughly a thousand times more distant than the Moon, to say nothing of the planets farther out in the solar system.

It is interesting to consider how long it would take to go to other stars. As we all remember from the Apollo program, which put the first men on the Moon, it takes about four days to go from Earth to the Moon. Traveling by spaceship from Earth to the planet Mars would take on the order of 260 days, one way. It is evident that our space travelers would get quite bored during that time, so they might make good use of their time by performing experiments involving quantum teleportation.

In order to get even farther out, we might use the accelerating force of other planets or even of Earth itself, as has been done with some of the unmanned spacecraft exploring outer planets. The idea is simply to have the spaceship pass close by a planet so that, by means of a sort of slingshot action, it can be accelerated into a new orbit that carries it much farther outward. For example, using these methods, the spacecraft *Pioneer 10* took about eleven years to travel past the outermost planets of the solar system on its probably unending journey into the space between the stars. We can thus estimate that it will, for example, take *Pioneer 10* about 100,000 years to get to Proxima Centauri, the closest star except for the Sun, at its current speed.

Perhaps, therefore, it would be good to have some other way to get around, to cover large distances. What we want is to travel anywhere instantly, without any limitation on how far we can go. Is that possible, at least in principle? This is why science-fiction writers invented teleportation. Magically, you disappear from one place, and, magically, you reappear at another place, just an instant later.

THE STUFF CALLED LIGHT

The first teleportation experiments were done with light, but what is light? Humans have always been fascinated by light. Probably long before we learned to write things down, people must have discussed how it is possible that through light, we experience objects close by or even at large distances. There are two basic concepts physicists use to explain how something travels from a light source—say, the Sun, or even a tiny candle—all the way to our eyes so we can recognize the object that emits the light. One concept assumes that light travels to us as *particles*, pieces of something, just like chunks of matter. The other assumes that light travels to us in the form of *waves*.

The simplest analog for the particle concept is that light travels just as a bullet or a small marble does. For the wave concept, the simplest pattern we can think of is the pattern of waves spreading out on the surface of water, for example, in a small pond. These two simple images convey the essential features of the particle and the wave concepts.

In the case of the marble, we have something localized—restricted in space—that moves. Similarly, the particle of light moves from place to place—from the light source, to the object we see, to our eye—by following some trajectory. Furthermore, just as marbles or bullets come one by one, the light source, for example the Sun, emits many tiny particles of light that travel toward us. They hit, for example, the tree across the road, some of them are reflected and scattered off that tree, and a few finally are collected by our eyes.

In contrast, the wave on the surface of a pond is not localized at all. If we throw a stone into a quiet pond, we see a wave that eventually spreads out all over the pond (Figure 1). Furthermore, waves do not

Figure 1. The nature of waves. Waves spread on a pond from the point where a stone was thrown into the water.

come in pieces, in chunks, but, rather, a wave can come in any size. There are very tiny waves caused, for example, by the legs of a small insect gliding across the quiet pond, or huge waves created by large stones thrown into the water. So there is continuity to the size of water waves.

So the big question is, What is light? Which concept applies to the phenomenon of light—the wave concept or the particle concept? Which of the features we just listed are actually features of light?

Much of the history of physics can be written as a history of the nature of light. Very early on, people started to carefully investigate which of the criteria for particles or for waves apply to light. In the early 1700s, there was a large battle between adherents of the particle picture, led by Isaac Newton, and followers of the wave picture, led by Robert Hooke. Back at that time, the particle picture triumphed. Many say that the weight and authority of Newton tipped the scales.

LIGHT IS A WAVE

In 1802, the English medical doctor Thomas Young performed an experiment that turned out to be crucial for our understanding of the nature of light. The experiment itself—actually one of the great experiments in the history of science—is extremely simple. Thomas Young just let light pass through two pinholes in a screen.

Behind the pinholes, he observed light and dark stripes (top sketch in Figure 2), called "interference fringes" today.

What happens if we cover one of the two slits? Then we do not see any fringes, but rather a broad patch of light (middle sketch in Figure 2). If we cover the other slit, we get a similar broad patch of light slightly shifted (bottom sketch). There is a large region where the two patches overlap.

From a particle-picture point of view, when we open both slits, we would expect that the light on the screen would be the sum of the two. But this assumption turns out to be wrong. Instead, in the overlap region, Young observed bright and dark stripes—the fringes. So there are positions, the dark fringes, where no light at all arrives when both slits are open. But when either slit is open alone, we have light there. Careful measurement shows that at the bright fringes, the amount of light is more than the sum of the two intensities that we would get with just either slit open. How can that be explained?

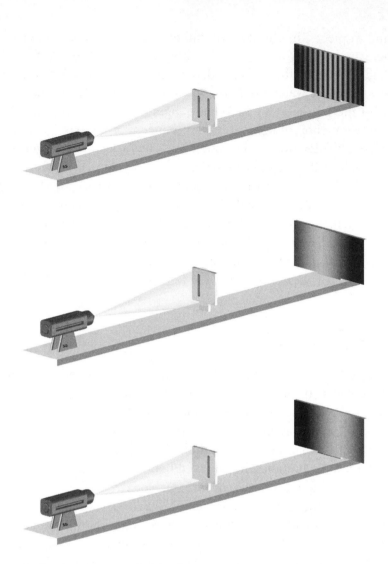

Figure 2. Thomas Young's double-slit experiment in a modern version. The light emitted by a laser passes through two slit openings in a diaphragm. Finally, it hits an observation screen. When both slits are open (*top*), we see a series of dark and bright stripes, called "interference fringes." If only one of the two slits is open (*middle* and *bottom*), we observe a broad illuminated area without any stripes. It is clear that the striped pattern in the top picture, when both slits are open, is not the sum of the two others. Rather, at the dark locations, the two waves coming there from the two slits extinguish each other. At the bright locations, they reinforce each other. The extinction at the dark stripes and the amplification in the bright stripes are a clear confirmation of the wave nature of light.

The wave picture provides an explanation of the fringes. Let's assume that a light wave comes from a certain direction, say, from the left, as shown in the figure. It hits the two-slit opening. On the other side of each slit a new wave starts. The two waves reach the observation screen. At the center line on that screen, the two paths leading from the slits will be of equal length. In that case, the two waves will oscillate in sync and they will mutually reinforce each other, and a bright stripe results. If we move our observation point, right or left in the figure, one of the paths gets a little shorter while the other one gets longer. The two paths leading from the two slits to any given point on the observation screen are no longer of equal length. There is a difference in path length.

So, depending on where exactly the new observation point is, the two waves will get more and more out of step. At some point, the two waves will be completely out of step. Where one wave is at its maximum, the other one is at its minimum. Where this happens, the two waves cancel each other out. Just consider the same situation for water waves. If two waves meet so that the crest of one meets the trough of the other one, they cancel each other out.

If we move even farther out, the path length difference will keep getting larger. At some point, the path length difference will be exactly one wavelength. In that case, crest meets crest again: the two waves reinforce each other and a bright stripe will be seen.

If we move the observation point even farther out, the pattern repeats. There will again be positions where crest meets trough: the waves cancel each other out, there is no light, and it will be completely dark, and so on. The interference fringes appear because in those places where we have mutual reinforcement, we get more light resulting in the bright fringes, and in those places where crests meet troughs we have the complete extinction of light—the dark fringes, destructive interference. So we see a striped pattern.

After Thomas Young's experiment, physicists no longer doubted that light consists of waves and not particles.

LIGHT IS PARTICLES

Then, in 1905, a completely unknown clerk at the Swiss patent office in Bern published a series of papers that changed the nature of physics. At that time, Albert Einstein was only twenty-six years old. In one of the papers, he proposed his relativity theory. But it is the first paper published in that year on which we focus now. It is the only one of his works that Einstein himself, in a letter to his friend Conrad Habicht, called "very revolutionary." In that paper, Einstein suddenly suggested that light is made of particles.

These particles of light, also called light *quanta*, later were named *photons* by the American chemist Gilbert Newton Lewis in 1926. In the face of all the evidence for the wave nature of light existing in Einstein's time, with the double-slit experiment being only one proof, how did this young clerk at the Swiss patent office in Bern dare to come up with the idea that light might be composed of particles, just the opposite concept? To discuss this question in detail, we have to learn something about the way physicists describe order and disorder.

SHEEPDOGS AND EINSTEIN'S PARTICLES OF LIGHT

There are many competitions worldwide every year to find out which sheepdog is the best. One of the jobs such dogs must perform is to gather a flock of sheep and move them to one specific place, say, into one corner of a field. From a physicist's point of view, what the sheepdog does is increase the order of the system. Before, the sheep might be scattered all over the field, particularly if they feel safe and no enemy is around. The sheepdog has something in its genes that tells it how to gather the sheep all together into one pack. In sheepdog competitions, that dog wins who herds the flock together in the shortest time, who gathers all the sheep in an orderly way at some place its master specifies.

Actually, the situation is very similar to clearing off the stuff—books, pieces of paper, and brochures—cluttering a desk. Most desks after some time look completely messy, a piece of paper lying here, a newspaper over there, a coffee cup on top of the newspaper, some other piece of paper in another corner, and so on.

Just as the sheepdog puts all the sheep into one corner of the field, one way to increase the order on the desk is simply to make, say, three stacks, one for notes, one for newspapers and journals, and one for books (Figure 3). Suddenly, all these items are in place and the rest of the desk is free. But unless we take care, the stuff will spread all over the desk again after some time. So, in both the case of the sheep and the case of the desk, we have a natural tendency for stuff to spread out evenly over the available space, and we also see that it takes a special effort to gather the stuff together again. The situation where the stuff is gathered together is one of higher order than the situation where it is evenly spread out.

Interestingly, a gas in a container behaves in the same way. Suppose

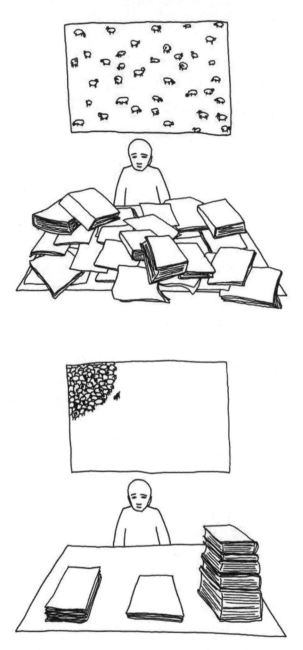

Figure 3. It always requires an effort to create order (*bottom*) out of disorder (*top*). In the case of sheep on the field, as shown in the picture on the wall, it is the effort of the sheepdog that provides the order. In the case of the mess on the desk, it's the effort of the person. Unfortunately, all systems in general have a natural tendency toward increasing their disorder.

we have a vessel separated into two parts by a wall with an opening that can be shut or not (Figure 4). We start with the valve closed, and all the gas particles are in one half, while the other half is empty. Then, we open the shutter. It's obvious what will happen. The gas will spread out evenly over the whole vessel. There will be a lower density then, since the gas has to thin out. Gas really consists of atoms and molecules. So the situation is just like two fields, one of them full of sheep. If we open a gate to the other, empty field, after some time, the sheep will spread out over both fields. We assume that there is an equal amount of food available on both sides and that there is no danger—and no sheepdog—pushing them either way.

Now, let us assume the opposite. Let's assume we start with the gas filling both sides. Will it ever happen that all the atoms will, of their own accord, move to one half of the vessel, leaving the other half empty? Probably not. Why not? In principle, such behavior is not impossible. If we look closely, we will observe atoms going through the opening in both directions, some from the right to the left, and some from the left to the right. It could happen by sheer coincidence that at some time, all the atoms are assembled on one side and none on the other. But apparently that is very unlikely.

Likewise, it is very unlikely that within one container, all the atoms will move into one corner together, leaving the rest of the container empty. In principle, this is not impossible; since every molecule moves around in its own zigzag fashion, it could simply happen by chance that they, at some instant, end up in one place. But this is extremely unlikely. The contrary obviously happens easily. If we have all the atoms in one corner and let them fly freely, they will immediately fill the available space in a homogeneous way.

This leads us to the observation that the universe tends to increase its disorder. All the atoms in one corner together means a highly ordered situation; all the atoms filling the available volume is a less ordered one. Also, we can see a clear connection between probability and order. The more orderly a situation is, the less probable it becomes.

Physicists like to describe disorder using the notion of *entropy*. Entropy is just a measure of disorder. More precisely, entropy reflects in how many ways a given situation can be realized. The larger the entropy, the more disordered a situation is, and the larger the entropy, the higher the probability for that situation to exist. For the case of gas filling a certain volume, it turns out that entropy increases with the in-

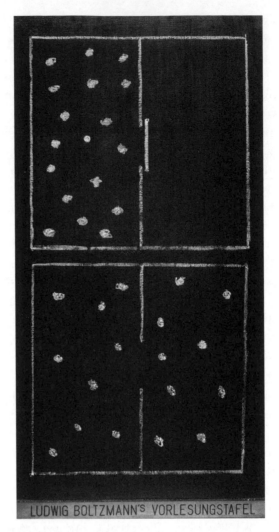

LUDWIG BOLTZMANN'S VORLESUNGSTAFEL

Figure 4. Gas, limited initially to one side of a container (*top*), immediately spreads out all over the place when the shutter is open (*bottom*). This condition will remain so, with the individual gas particles going back and forth between both sides. The reverse procedure, where all evenly dispersed particles of a gas (*bottom*) assemble on their own on one side only (*top*), never happens. The reason is that this latter procedure is, statistically, highly improbable. It would mean that all molecules happen by chance to move the same way across the opening. Albert Einstein observed that light inside a cavity behaves in a similar way and therefore assumed that light, like gas, must also consist of particles. The connection between probability and order was discovered by the Austrian physicist Ludwig Boltzmann. His blackboard ("Vorlesungstafel") was used to draw the sketches for this book.

crease in volume. In a sense, putting all the atoms in the smaller volume is like putting them all together in a corner.

In 1905 the young Albert Einstein made an important discovery. He studied the entropy of a gas filling a certain volume and compared it with the entropy of light filling the same kind of volume. What Einstein discovered was an interesting coincidence. He read and compared scientific papers that had already been around for more than five years, so anyone else could have made the same discovery. But Einstein discovered that the entropy—remember, this is the measure for disorder—of radiation in general, or light more specifically, filling a certain volume is very similar to the entropy of gas filling that same volume. Indeed, he found out that the mathematical expression derived by the German physicist Wilhelm Wien for the entropy of radiation filling some volume is the same as the mathematical expression derived earlier by the Austrian physicist Ludwig Boltzmann for a gas filling a container. More precisely, the two mathematical expressions for entropy vary in the same way if one changes the available volume.

Einstein, seeing this analogy, made a very bold conjecture. He knew that the expression for the entropy of gas filling a volume can easily be understood on the basis of how the individual molecules move around, and how improbable it is that they fill only a part of the available space. Because of the analogy of the two mathematical expressions for light and for particles, Einstein assumed that light is also made up of particles, which move around just like molecules and which also don't like to end up in some corner if a large volume is available to them.

Einstein was very careful. He called his idea just a heuristic aspect in the title of the paper: "On a Heuristic Point of View about the Creation and Conversion of Light." We talk of *heuristics* when we talk about an idea that helps us to find out, to make guesses, to get a feeling for a situation. We do not necessarily imply that we are able to prove our point. Maybe Einstein did not want to offend the adherents of the wave theory of light too much. But in the paper, Einstein himself is very explicit. In his 1905 paper, he takes the particles of light quite literally, as localized points moving around in space, just like atoms.

Einstein was not content with just making this bold guess. He thought about where else in nature the idea of light being particles would also have interesting consequences. He suggested that his explanation, if correct, would also help to explain a phenomenon not understood by physicists at his time—the *photoelectric effect*.

The German physicist Wilhelm Hallwachs had discovered this effect in 1888. He saw that something interesting happens when we shine light on a metal plate. The light can actually free electrons out of the metal, and these electrons can easily be detected as an electrical current. People had tried to explain that effect using the wave concept of light, which was, at that time, the accepted one. But the wave idea ran into serious problems when it came to explaining what experimentalists saw.

One of the unaccountable features is that the electrons start to appear instantly when we start to shine light on the metal plate.

Why is that a problem for the wave concept? Well, a wave is an oscillation back and forth. In the case of light, it is an oscillation of an electric and a magnetic field. When light hits a metal plate, it will make the electrons inside the metal plate start to oscillate. The electrons would oscillate initially just a little bit, and they would oscillate more and more as they absorbed more and more light, until finally they would break loose from the surface of the metal and be free. Imagine a person on roller skates who oscillates back and forth in a half tube. By pumping with his feet, he pumps more and more energy into the oscillation until he is able to jump above the tube. Evidently, it takes a while to accumulate enough energy to do this. So, for our photoelectric effect, it would not be so surprising if the electrons started coming out right away when we illuminated our metal surface with a strong light beam, because then the electrons could be brought to large oscillations in a very short time. But people did experiments with very weak light and found out that the electrons start coming out immediately anyway. According to the wave concept, it should take some time until the electrons have accumulated enough energy. This was just one problem with the wave concept of light.

However, if light consists of particles, this problem is resolved. What happens, as Einstein remarked, is simply that an individual light particle, an individual photon, just might, by chance, kick out an individual electron (Figure 5). This explains why the electrons start coming out immediately after we turn on the light beam, and it also explains why the amount of electrons observed is strictly proportional to the amount of light shining on the surface. Double the light intensity means double the photons hitting the surface, which again means double the electrons set free.

Einstein made another bold prediction based on this model, and this is where physicists always prove their worth. The final test of a the-

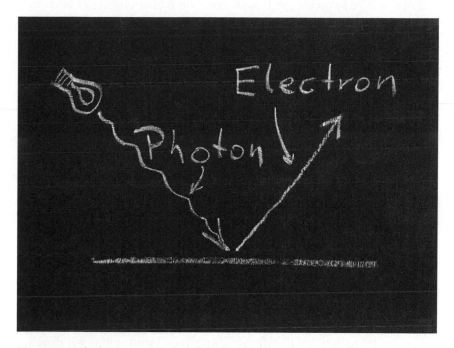

Figure 5. Light hitting a metal surface can kick out electrons, which then move away. This is the photoelectric effect. Einstein explained it by saying that light consists of individual particles, called photons.

ory is not just that it explains some phenomena already observed in the laboratory or in nature. The most convincing argument for a new theory is that it is able to predict something no one had been able to calculate so far and something that had not been observed until then. What Einstein predicted for the photoelectric effect is a connection between the frequency of the light shining on the metal plate and the energy of the electrons coming out.

Suppose a photon with some energy hits the metal plate. It might kick out an electron or it might not. Suppose it does. So, how fast will the electron move after it has left the metal plate? What will its energy be? There are many things that can happen. The photon might hit the electron such that it cannot convey all its energy to the electron, just as a billiard ball might hit another one and still keep moving. But it could happen that the photon would hit the electron such that it transfers all its energy to the electron. Then, the electron would move quite fast. But it might lose some of its energy inside the metal plate before it gets out. But it could also happen that the electron would be hit by the photon right on the surface of the metal such that the electron would not lose any energy before it got free. But it still would have to get out. Every surface has some attraction to electrons. This amount of attraction depends on the material of the surface, but the electron always needs some energy to overcome this attraction.

So, what this all boils down to is the following. If we get lucky and the photon hits the electron in the right way, and if we get luckier still and the electron somehow does not lose all its energy inside the metal, the electron comes out with the original energy of the photon, reduced by the energy required to get out of the metal. Now we have to ask ourselves, What is the original energy of the photon? And there, Einstein took an idea suggested by Max Planck five years before, that is, that the energy is quantized. It comes in chunks, in multiples of quanta such that the energy of a photon E is given by its frequency v multiplied by a constant, Planck's quantum of action h: $E = hv$. The most important point now is this: if Einstein is right, then the maximum energy with which the electrons come out of a metal must be increasing in proportion to the frequency of the light shining on the metal plate. This prediction of Einstein's was confirmed in a very beautiful experiment by the American physicist R. A. Millikan in 1916. For his work on the photoelectric effect, Einstein was awarded the Nobel Prize in Physics in 1921.

EINSTEIN AND HIS NOBEL PRIZE

Actually, the story of Einstein's Nobel Prize is quite interesting. In general, in order to be awarded a Nobel Prize, a person has to be nominated. The Royal Swedish Academy of Sciences, which chooses the prize recipient every year, has a small selection committee. Many physicists all over the world are invited to make nominations, and the academy then selects out of these nominations the laureate, or up to at most three joint laureates. For the final decision, the academy draws upon recognized experts in the respective fields and asks them to provide their opinion about the nominations. In Einstein's time, this expert work was done mostly by the members of the Nobel committee themselves.

Einstein had been nominated many times, for the first time in 1910, only five years after his annus mirabilis. In that notable year of 1905, Einstein wrote five scientific papers. In one of them he proposes his relativity theory. In another one, the most famous equation in physics, $E = mc^2$, appears for the first time. In the third one, in the field of atomic physics, he gives a very good estimate of the size of atoms. But the first paper that was published in 1905 is the one where he proposed the concept of photons.

Nearly all the nominations that Einstein received were for his relativity theory. The problem was that there were two members of the Nobel committee who either did not like the theory of relativity or even thought it was wrong. The fact that Einstein had not received the prize started to become a strange situation in the scientific community. The breakthrough came when the theoretical physicist Carl Wilhelm Oseen became a member of the academy and realized why Einstein had not received the Nobel Prize. Oseen suggested giving the prize to Einstein "for his discovery of the law of the photoelectric effect"—in essence, for his concept of particles of light, of photons. Finally, Oseen was able to convince the committee; it decided to give the Nobel Prize for the year 1921, which had not been awarded, to Einstein in the year 1922, "for his services to theoretical physics, in particular, for his discovery of the law of the photoelectric effect." Interestingly, the academy also issued a cautionary statement, which shows that some of the academy members in 1922 still felt that there might be some danger that the theory of relativity could turn out to be wrong.

A question still open today is why Einstein never got the Nobel Prize for the relativity theory later on. He lived until 1955, and there have been cases where people received two Nobel Prizes, even two Nobel Prizes in Physics. But it might very well be that Einstein was just never nominated again. Today, relativity theory has even found widespread technical application. The worldwide global positioning system (GPS) would not work properly if it did not take into account the fact that the precision atomic clocks it uses on satellites run at a different speed than they would on the ground, a consequence of Einstein's theory of relativity.

A CONFLICT

We just saw that light may be explained in two different ways, either as a particle or as a wave. We saw that there actually are specific experiments that confirm either view. From Young's experiment, we can conclude that light is a wave. But the photoelectric effect seems to confirm that light consists of particles. We might not worry much about this, but we have also learned that particles and waves are two completely different concepts. So there is apparently a conflict. Which idea should we believe?

Einstein was aware of this conflict, and he actually mentions something along those lines already in his first 1905 paper. This conflict might very well be the reason why Einstein called that paper "very revolutionary." He was forced to put aside all the experiments that confirm that light is a wave, to disregard them, when he suggested that light is a particle. Maybe today, more than a hundred years later, we are able to answer the question in a better way. Is light a particle or a wave? Is there any way to understand, for example, Young's experiment from the position that light consists of particles?

It was clear to Einstein that the interference phenomena that are so characteristic of the wave concept and are apparent in Young's double-slit experiment cannot be brought into agreement with the particle concept. Let's consider that experiment once again (see Figure 2). When both slits are open, we see dark and bright stripes on the observation screen. In terms of the wave concept, these stripes are easily explained. When a wave passes through both slits, it gives rise to two separate waves, each emerging from one of the slits. These partial waves move toward the observation screen. At some locations, the two waves oscillate in sync—that is, they oscillate the same way, up and down—so they reinforce each other, which gives rise to the bright stripes. At other places on the observation screen, the two waves oscillate in completely opposite ways. While one goes up, the other one goes down. So, they cancel each other out, and this gives rise to the dark fringes, to the dark stripes. Now we ask the following question: Is there any chance of understanding this phenomenon if we consider that these are not waves that pass through the two slits, but particles? Here's the issue: a wave can fill a large region of space. It can, in our case, pass through both slits. A particle has to decide whether it goes through this slit or through the other one.

A deeper understanding of how to reconcile the particle picture and the wave picture of light had to wait until after the development of the full quantum theory by the German physicist Werner Heisenberg and the Austrian physicist Erwin Schrödinger in 1925–26.

HOW WE BECAME CERTAIN
OF UNCERTAINTY

It is the year 1925. A young German physicist named Werner Heisenberg—only twenty-five years old at the time—had just finished his Ph.D. and was working at the University of Göttingen in Germany. Over the years, Göttingen had become one of the world's centers of science, where many of the great new ideas in mathematics and physics originated. Evidently, this was what had attracted young Heisenberg also, who had done his studies at the Ludwig-Maximilians-Universität, the University of Munich.

Heisenberg decided to work on the most important outstanding problem in physics at that time—understanding atomic physics and its connection to the idea of the quantum. Let's recall the situation—and it was a desperate one that had already lasted for a while. In 1900 Max Planck had introduced the concept of the quantum purely as a mathematical trick. He needed it to explain the specific color a glowing body has at a certain temperature—more precisely, the specific distribution of colors emitted. Planck's act of desperation was not accepted and was even rejected by many physicists, including himself.

The person who then jumped onto the idea and really brought it to life was Einstein, in 1905. As we have seen above, he was bold enough to not just view Planck's idea as a mathematical trick, but to assume that light is really made up of quanta, of individual particles. That way, he was able to very simply explain the photoelectric effect—the way in which electrons are kicked out by light from a metallic surface—an outstanding puzzle at that time.

Following that, in 1913 the Danish physicist Niels Bohr used the quantum to build up his model of the atom, which was very similar to

the model of the planets moving around the Sun. With all these suc-
cesses, what then was the problem?

The quantum was a very successful idea that apparently explained
a number of phenomena. The problem in 1925 was that there was no
full mathematical theory incorporating the quantum idea. The works of
Planck, Einstein, and Bohr were more or less ways of playing around
with the new concept and finding a successful application. But the
challenge was to find the fundamental mathematical equations from
which all this followed. It was obvious that there was something very
deep hidden in the quantum idea, and it was also clear to everyone that
this something had not been found yet. Scientists in general and physicists in particular are very immodest. They are only satisfied if they find
a profound explanation. Explanation in physics always means first finding a mathematical equation that describes the phenomena observed.
And that's not enough. Physicists are not satisfied with finding a mathematical equation that describes a specific phenomenon. They want to
dig deeper—they want to find a reason for the validity of the equation.
They want to find fundamental laws.

It is exactly the mathematical description of quantum phenomena
that was lacking in 1925. People had been working very seriously for a
long time on this. Young Heisenberg decided to find out what kind of
mathematical laws were needed. He was unable to find a solution, and
in Göttingen there were too many distractions. Luckily for physics,
young Heisenberg developed a serious bout of hay fever.

That attack must have been very serious, because his professor, Max
Born, sent him for a couple of weeks to the island of Helgoland in the
North Sea. Helgoland had been known for a long time as an ideal resort
for hay fever sufferers.

There, Heisenberg had lots of time to ponder this mathematical problem. The story goes that Heisenberg took many long walks
through the woods and along the coast. Quite suddenly, the right idea
occurred to him, and he discovered a new mathematical structure that
provided him with the fundamental laws of quantum physics. This great
discovery won Heisenberg the Nobel Prize in Physics in 1932, and it
instantly made him, at twenty-five, one of the central figures in modern
physics.

These fundamental laws of quantum physics provided a way to
calculate what atoms actually do—in particular, to calculate what kind

of light an atom emits, as well as how the motion of electrons around the nucleus of the atom can be calculated, and many other things. But there was a drawback. There was a price to be paid, and the price was very high.

The basic point is that to observe a particle, we have to interact somehow with it. For example, we have to shine a light on an electron to see where it is. As we saw when we discussed Einstein's explanation of the photoelectric effect, this results in a kick to the electron. After the kick, the electron moves at a different speed than it did before. So, the speed that we determine now will simply be wrong. But wait a minute. We should be able to account for that, we think. By finding out how large the kick was, shouldn't we be able to calculate backward and find out what the speed of the electron was initially? As it turns out, this cannot be done if light consists of individual particles, quanta of light. These quanta of light, when they kick our electron, might randomly fly away into any direction. And depending on the direction, the kick will impart a different change of momentum, that is, an uncontrollable modification of the speed of our electron.

It gets even worse. If we want to find out precisely where the electron is, we need to use shorter and shorter wavelength light. It turns out—and this was confirmed by experiment—that the shorter the wavelength we use, the larger the momentum kick. That is, as the uncontrollable disturbance of the electron becomes larger, we can more precisely determine the electron's position.

Now, we see, there is some sort of payoff. The smaller the wavelength of the light, the more precisely we are able to determine the *position* of the electron. But this implies at the same time that the kicks get larger and, therefore, we can less precisely determine the *speed* at which our electron moved initially.

This is exactly Heisenberg's uncertainty idea, which he stated in a more mathematical way. It says simply that we cannot know both the position of any object and its momentum (the magnitude of its speed multiplied by its mass) at the same time with arbitrarily high precision. If we know the position very well—that is, if its uncertainty is very small—we know less about how fast the object moves, and vice versa. That uncertainty can be very large. Mathematically, this is expressed by the famous Heisenberg uncertainty principle.

Momentum and position, because of Heisenberg's uncertainty

principle, are said to be complementary to each other. This notion of complementarity was introduced by Niels Bohr as one of the great lessons we learn from quantum physics. Simply stated, we are unable to know the world with complete precision. We always have to make our choice.

QUANTUM UNCERTAINTY:

JUST OUR IGNORANCE, OR IS IT THE WAY THINGS ARE?

Heisenberg's uncertainty principle has immediate consequences for understanding atoms. The atom consists of an atomic nucleus with electrons buzzing around it. An atom has the size of about 10^{-10} meters. This is a tenth of a billionth of a meter. Somewhere in the atom is an individual electron, on which we now focus our attention. Let's assume we don't know more than the fact that the electron is confined to these 10^{-10} meters. That means we are uncertain about the position of the electron. Its position uncertainty is about 10^{-10} meters.

What does the position's uncertainty imply for the uncertainty with which the speed of the electron is known? Heisenberg's uncertainty principle says that the product of the momentum uncertainty and the position uncertainty cannot be smaller than a certain size given by the quantum of action discovered by Planck. Simple calculation using Heisenberg's uncertainty principle tells us that the speed of our electron in the atom is uncertain to a very large extent. The uncertainty is actually of the order of 1,000 kilometers per second. And this is only the uncertainty. This shows just how little we know about the speed, if we happen to know that the electron is inside the atom.

On the other hand, we know that the electron moves around there, buzzing back and forth. So, its speed averaged over a few cycles where the electron goes around is zero, since the electron still stays with the atom. In other words, if we look at the atom at a certain time and then again, much later, we still find the electron inside the atom, and the buzzing back and forth of the electron itself has averaged out. So, if the average speed is zero and the uncertainty is of the order of 1,000 kilometers per second, this means that the electron moves actually with speeds of up to a few thousand kilometers per second. This is indeed confirmed by experiment.

But wait a minute. Aren't we confusing two things right now? Our argument was that we cannot know position and momentum at the same time because measurement of one disturbs the other. But it could still be the case that the electron is at a specific position, even if it might be unknown to us, at any particular time and that it has a well-defined speed, say, 7,350 kilometers per second, at the very moment when it passes through that specific location. So it could very well be the case that every particle in the universe at some time is at some place and moves with some well-defined speed, but we just don't know what the particulars are. From that viewpoint, Heisenberg's uncertainty principle would simply tell us that we can never find out both the position and the speed of an electron because of the unavoidable disturbance of the observed system caused by the measurement.

Just for fun, let's consider briefly what this uncertainty would imply for cars. Suppose Heisenberg's uncertainty principle applied to cars and had an important impact there. Then, the following dialogue (Figure 6) would be possible.

THE QUANTUM EXCUSE

POLICE OFFICER (*stopping a car for speeding*): Sir, I just measured with my radar that you were moving with a speed of 40 miles per hour. That is too fast. Do you know what the speed limit is here?

DRIVER: Yes, 30 miles per hour, just like everywhere in the city.

POLICE OFFICER: So, why did you surpass the speed limit? Didn't you know that you were going too fast?

DRIVER: No, I had no idea at all that I was going too fast.

POLICE OFFICER: Aren't you watching your speedometer? You should keep an eye on it.

DRIVER: There's no point in keeping an eye on my speedometer.

POLICE OFFICER: Why is that? It's your obligation to obey the speed limit.

DRIVER: I decided to disconnect the speedometer.

POLICE OFFICER: Are you joking? That's illegal. You're tampering with a device that is important to your driving safety. I could actually have your car removed from the road now, until the speedometer is fixed.

Figure 6. "I decided to disconnect the speedometer."

DRIVER: Please don't do that. I had a very good reason for disconnecting the speedometer.

POLICE OFFICER: What could be a good reason? Maybe you just want to close your eyes and not worry about the law.

DRIVER: No. I recently read a very good popular book about quantum physics, and that stuff is really mind-boggling.

POLICE OFFICER: OK. Sure. But does the book tell you to disconnect the speedometer?

DRIVER: Not directly, but there was this guy Heisenberg . . .

POLICE OFFICER: Oh, the one with the uncertainty principle?

DRIVER: Yes! He told us that we cannot know the position of the car and its momentum at the same time. In other words, I cannot know both where I am and how fast I am going. So, since it's very important for me to know where I am, I decided to disconnect the speedometer.

POLICE OFFICER: But disconnecting the speedometer just means that you don't know how fast you go. The car is still going at a certain speed, as I just found out!

DRIVER: I've learned that we shouldn't assign properties to objects unless they are measured, and if no one measures how fast the car goes, that is fine.

POLICE OFFICER: But I measured it, and I saw that you were going 40 miles per hour.

DRIVER: You're correct. You observed me going 40 miles per hour, but that does not mean I was going that fast before you watched me. It only means that there was some probability that I could be observed going that fast.

POLICE OFFICER: Don't pull my leg. Are you now implying that it's my fault that I found you going that fast? This might become very expensive for you. Just pay your fine and go ahead.

DRIVER (mumbling to himself quietly): Typical of the police. Instead of admitting it when they lose an argument, they just play the powerful representative of the law. (Loudly) OK, officer, I'll pay, but I'm still worried about reconnecting the speedometer.

POLICE OFFICER: Did you put in any numbers? Did you ever find out what the Heisenberg uncertainty principle means for your car?

DRIVER: No, I didn't, but I learned that it applies to everything.

POLICE OFFICER: OK. I'll charge you for speeding, but not for having disconnected the speedometer if you promise me you'll figure out what the Heisenberg uncertainty principle means for your car.

So what does the Heisenberg uncertainty principle imply for a car? Suppose you want to know the position of a car weighing about one ton to within, say, one millimeter. So the position uncertainty is one millimeter. The Heisenberg uncertainty principle implies that the uncertainty of the speed of the car is only of the order of 10^{-34} meters per second. So, if the police officer measured the car's speed at 40 miles per hour, he made only a tiny mistake that we can safely neglect. According to Heisenberg, the larger the mass of the object, the smaller the uncertainty in speed. Therefore, since the mass of the car is so large, the uncertainty in the speed is quite small. But for small objects with very little mass, like an electron, this effect is not negligible at all.

Let's return to our discussion of what the Heisenberg uncertainty principle really means. To assume that it is just an expression of our ignorance seems to be most natural. What could be wrong with particles having their exact positions and exact velocities at all times, and our just not being able to measure both at the same time? Is there an alternative? What could that alternative look like? It turns out that the point of view we just discussed was the point of view of classical physics until the invention of quantum mechanics, and it still is the working assumption of many scientists today. Einstein is one of the most prominent advocates of this, as it is called, "realist" position. From that point of view, Heisenberg's uncertainty principle is just an expression of the limits of what can be determined by measurement. Or in philosophers' terms, the nature of uncertainty would be an epistemic one. Epistemology is that part of philosophy that deals with what we can know and how we might come to know what we know.

The alternative philosophical position is to assume that the uncertainty principle is not simply a statement about what we can know; it is a statement about the nature of things. From that point of view, Heisenberg's uncertainty principle is a statement about how things are and what features they have. It's a statement about what exists. A philosopher would call such a position about the nature of the uncertainty principle an ontological one. From that point of view, the electron would have neither a position that is better defined than the position uncertainty tells us, nor a speed that is better defined than its momentum uncertainty. That ontological position was held by Bohr.

So, is quantum uncertainty epistemic or ontological?

What does it mean to assume that the uncertainty principle is not just a limit to what we can know, but that it describes how things actually are? It would mean that an electron never *has* both a well-defined position and a well-defined momentum at the same time. The electron neither *is* at a certain place, nor does it move with a specific speed. In a sense, the electron carries the possibilities of many velocities at the same time and the possibilities of being in many places at the same time. How can that be? How can we make sense out of that? Can an electron move at a number of speeds at the same time? If it moves at a high speed and at a low speed together, wouldn't that mean that the electron would break apart?

How can we account for the uncertainty, say, in momentum — that is, in the velocity of the electron — so as to claim that the electron actually carries the possibilities for a range of velocities at the same time? This does not make sense as long as we stick to the concept of the electron being a point moving around. We have to go to another picture. This idea was brought to us by the French physicist Louis de Broglie, who introduced the wave nature of matter. According to that concept, every particle moving at some speed also has a wave associated with it, the wavelength of which corresponds to its speed. The larger the speed, the smaller the wavelength. If we accept that description, our task of assigning an electron a couple of speeds at the same time becomes easy. All we have to do is to take the bridge over to the wavelength and assign an electron a couple of waves with different wavelengths. The wave making up the electron is then obtained by adding all these partial waves with different wavelengths together. And then, it turns out that the uncertainty in momentum, that is, the uncertainty in speed, simply implies that we have to superimpose waves within a wavelength range. This makes up a wave packet (Figure 7). It turns out that if we add waves within a certain wavelength band the right way, then a wave packet that is localized results. That means that the individual waves, which might be very widely extended, extinguish one another, except in a small region. The picture then is that the small wave packet corresponds to the particle. Actually, we have to ask ourselves now, is the particle a point or is it extended? Suppose the particle is a point that is smaller than the size of the wave packet. What does our picture mean?

The interpretation is this. The wave packet is not an object like a car or a tennis ball. The only function of the wave packet is to give us the probability of finding the particle somewhere, should we perform a measurement of its position. Far out, where the wave packet is effec-

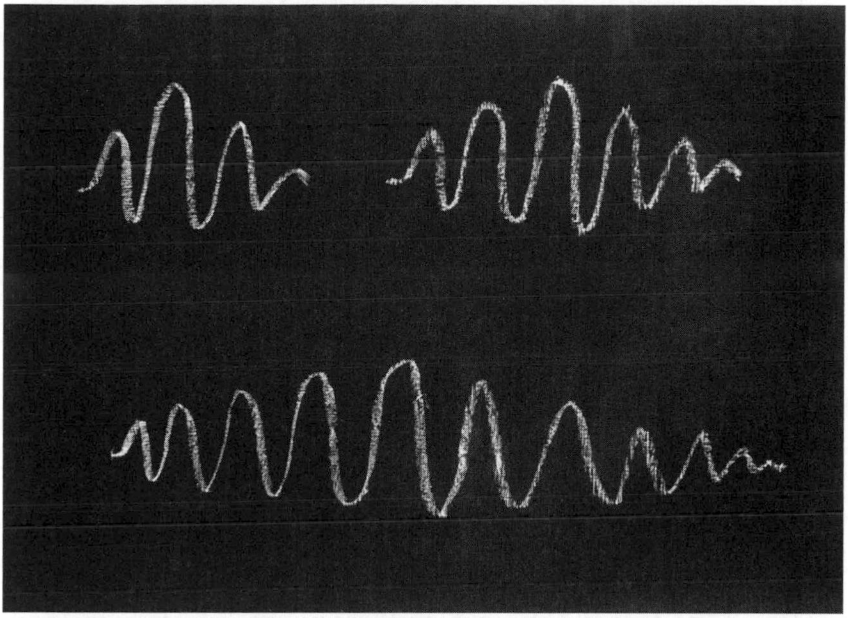

Figure 7. Various wave packets, which are built up by adding individual waves of different wavelengths together. Short wave packets (*top*) mean a small position uncertainty, because the electron has to be somewhere within the wave. They are built of individual waves with a large wavelength spread. This means a large momentum uncertainty, that is, a large uncertainty about how fast an electron moves. Long wave packets (*bottom*) mean a large position uncertainty, since we do not know where within the wave the electron is. The momentum, and therefore the speed, of the particle is then better defined because such packets are composed of waves with a smaller wavelength speed.

tively zero, the particle cannot be found. The probability is negligible that the particle can be found out there. In the center of the wave packet, at the maximum, we have the highest likelihood of finding it. The wave packet concept implies that we cannot be certain of where to find the particle. Actually, we can find it anywhere within the wave packet, and when we do a specific measurement for a specific particle, it is just pure chance where exactly we find the particle within the wave packet.

We also arrive immediately at an interesting consequence. It turns out that the more waves with different wavelengths we add up, the shorter the wave packet will be. But the size of the wave packet just corresponds to our ignorance of the position of the particle, because it could be found anywhere within the wave packet. Therefore, the shorter the wave packet, the better defined the position is. But each wavelength corresponds to one specific momentum or speed. Then the momentum, that is, speed, is less well-defined, because more different wavelength components are needed to make a short wave packet. This is the wave meaning of the uncertainty principle.

Experimentally, we can actually prove that the picture of the wave packet is correct. People have put together wave packets of varying composition, and they have found that for very short wave packets—that is, those with good position definition or small position uncertainty—they have to put together many different wavelengths, or momenta. On the other hand, people have shown that if we select a very narrow momentum band, the wave packet and thus the position uncertainty becomes very large. This actually holds for individual particles.

This very important point needs to be stressed again and again. The wave packet is something that we have to associate with each individual electron. So, the electron does not carry a precise momentum, that is, a well-defined speed, nor is it localized at a well-defined position. The wave packet means that the electron carries neither a certain velocity nor a certain position. Should we decide to measure the position, we will find the electron somewhere within the wave packet. The uncertainty of the electron's position suddenly has become much smaller. Physicists say that the electron has become localized because of the measurement. So the experiment does not reveal a position that the electron had before the measurement. All we had before was the wave packet, which prescribes simply where the electron could be found with a certain likelihood, or probability, but nothing more. Likewise, the wave packet consists of a number of different waves, each one corresponding to a certain velocity. If we perform a measurement of the particle's velocity, one velocity—that is, one specific wavelength—will result. But again, the electron did not have that velocity before. The essential point now is that Heisenberg's uncertainty principle is a statement about the nature of things, and not just about what we can know about the world.

THE QUANTUM VERDICT AGAINST TELEPORTATION

In science fiction, the teleportation procedure is usually the following. In the first step, the original is precisely scanned to determine all its properties. The scanner determines the states of all the atoms, all the electrons, all the particles inside the original object. This is a huge amount of information. In the second step, that information is sent over to the receiving station. The original is finally reconstituted using some matter. This could be some material that is already present at the receiving station, or the material could also be sent over, which is a cumbersome and unnecessary procedure.

The important point is that we have to find out the state of each particle that makes up the object. And we must send the information about the complete state to the receiving station so that the original can be reconstituted.

But hold on! What kind of measurement do we have to perform now? In general, we do not know the state of, say, a specific electron. What can we do to measure the state of the electron? We can decide to measure its position, or we can decide to measure its momentum, or maybe something else. The problem is that, by one measurement, we are not able to determine what the complete state is. Suppose we decide to do a position measurement. Then we are actually localizing the electron at some position, so we are changing its state. The electron's state after the measurement is different from its state before that measurement. In general, any measurement changes the state, and any measurement can give us only partial information about the state. The measurement itself destroys much of the information present before.

So we have no way to determine the unknown state of the electron by measuring it. We arrive, therefore, at the very important conclusion that it is not possible to determine the unknown state of an individual system. In other words, it is in principle impossible to obtain the full information that characterizes the object we want to teleport.

We therefore conclude that the procedure for teleportation as featured in science-fiction literature and movies can never work. This is a consequence of Heisenberg's uncertainty principle. It is impossible to determine the state of a space traveler who wants to be teleported or whatever else we want to send around. Heisenberg's uncertainty principle prohibits us from obtaining complete information about any individual system. So, unlike what our science-fiction authors imagine, it is impossible to start teleportation by scanning all the features of an object in order to send it to the receiving station.

Besides the fact that teleportation does not work the way science-fiction authors imagine, we have learned something much more important. We have learned something whose relevance goes far beyond teleportation and science fiction. We have learned that we can never know the world completely.

Let us step back for a moment and consider the scientific enterprise itself. We have literally millions of scientists all over the world. What are they doing? They want to *know* facts. They want to learn something about the universe. They want to find out the laws of nature. They want to measure things. They want to learn the individual properties of systems—electrons or elephants or whatever—and they want to *explain*. Modern science has done this in a particular way, and that path has been a golden road to success over the centuries. That way was to look closer and closer, to look more and more into details.

By looking deeper and deeper into the nature of matter, people found out all kinds of interesting and beautiful things. They found out that atoms are the fundamental constituents of matter. Then they found out that the atoms themselves are built up of electrons, protons, and neutrons. Then they found out that there are even more basic constituents, *quarks*. And the search is by far not over yet. Witness for example the impressive new experiments at CERN, the European Laboratory in Geneva, Switzerland.

But quantum mechanics suddenly tells us that there is a fundamental limit to completely know the state of the world. We have found out that there is a limit to knowing the states of individual systems and, therefore, the features of the world. It is not possible to determine the quantum state of any electron, or of any particle for that matter, completely. And since quantum mechanics is universally valid, at least as far as we know today, this holds for any object whatsoever. For all practical purposes Heisenberg's uncertainty principle can safely be neglected for objects in everyday life. We saw this with the small story where the car was stopped by the police for speeding. But someday, we will actually be able to demonstrate that quantum uncertainty has its relevance also for macroscopic objects. This is a question of technology as it develops. There is no clue in sight telling us that quantum uncertainty must stop somewhere.

Apparently, the story goes that the people who wrote for *Star Trek* learned one day about the limits imposed by Heisenberg's uncertainty principle. Probably one of their many fans in the scientific community told them. All *Star Trek* fans know how the filmmakers got around this. They invented the "Heisenberg compensator." This fictional device fixes the problem described by Heisenberg's uncertainty principle. Such a device is not possible for very fundamental reasons. Therefore, the mechanisms used in the Heisenberg compensator must remain unexplained. The story goes that Michael Okuda, the technical advisor of the *Star Trek* series, was once asked by *Time* magazine, "And how does the Heisenberg compensator work?" He replied, "It works very well, thank you."

So, what we have learned so far is that quantum mechanics ends the dream of teleportation. But the reader can take hope from the fact that there is a whole book in front of him or her about quantum teleportation, so there must be a way around these restrictions.

QUANTUM ENTANGLEMENT
COMES TO THE RESCUE

The solution is somewhat surprising, but in general not that new: in medicine, we have known for a long time, at least since the days of the medieval Dr. Paracelsus, that something that is bad and unhealthy sometimes can actually become its own cure if it is used in a different way. Translated into our situation, this means that quantum mechanics itself is called upon to come to the rescue.

To see how, let's look closely at what Heisenberg's uncertainty principle really tells us. It tells us that it is not possible to determine by measurement all the information necessary to characterize any individual system.

But do we need that in teleportation? Do we really need to determine all the information characterizing a system in order to teleport it? To answer that question, we step back for a moment and ask ourselves what we really want to achieve in teleportation. It is not important for us to *determine* all the information carried by a system. Precisely speaking, it would be enough to just *transfer* all the information over to the receiving station. Since there is no need to actually *know* the information, there is no necessity to measure it. Just ship the information over. Such a procedure would not be limited by the Heisenberg uncertainty consideration.

All we need is an information channel, a kind of tunnel by which the information can get directly from point A to point B. Measurement is not necessary. Actually, measurement should not be part of it. More precisely speaking, measurement of the information carried by the system *must not* be part of the procedure because measurement would bring in the limitations imposed by Heisenberg's uncertainty principle.

And we will see now that transferring information from A to B without measuring it is possible. That's the clue to the solution. The information channel is given by quantum mechanics through the entanglement phenomenon that Albert Einstein called "spooky." It is quantum entanglement that makes quantum teleportation possible.

The quantum mechanical solution was proposed in 1993 by an international collaboration of six theoretical physicists: Charles Bennett of IBM; Gilles Brassard, Claude Crépeau, and Richard Jozsa of the University of Montreal; Asher Peres of the Technion (the Israel Institute of Technology in Haifa); and William K. Wootters of Williams College. Actually, the affiliations are an interesting indication of international collaboration. While such collaborations had existed earlier, they became particularly easy with the invention of the Internet. While in the old days one had to write letters and wait awhile until answers arrived, today the Internet enables people in widely distant places to easily collaborate, discuss their ideas, make new suggestions, and together write a scientific paper much faster than before. The Bennett-Brassard-Crépeau-Jozsa-Peres-Wootters paper has the title "Teleporting an Unknown Quantum State via Dual Classical and Einstein-Podolsky-Rosen Channels."

To have the word "teleporting" in the title of a physics paper was quite unusual at that time, since teleportation was considered to be part of science fiction and a somewhat shaky topic. But apparently, there was no better name for the interesting theoretical discovery these people made, and it was a very fitting name indeed. In order to achieve teleportation, the authors suggest that we use "dual classical and quantum channels." In other words, we need a classical channel for communication and a quantum channel for the teleportation. Evidently, the quantum channel does the trick.

Let's first discuss what a classical channel is. Whenever information is sent from A to B—let's call the two positions Alice and Bob—one needs a channel that connects the two. A simple classical channel is a telephone line, and the information then travels along that channel from Alice to Bob and from Bob to Alice. In modern telephones, the information is digital. That is, it is a sequence of bits, each bit being either 0 or 1. At the sender side (Alice), the spoken word is converted into a stream of bit sequences that is sent over to the receiver (Bob), where the bit sequences are again converted, into sound.

The important point for our discussion is a rather obvious feature of the classical channel. The information is fed into one end of the channel, and it comes out at the other end. So the information must be there from the beginning; it must be well-defined. Otherwise, there is noise in the communication. Another point is that one can actually follow how the information travels along the line since it moves from point to point. Such a channel would not work for teleportation, since it is not possible to extract or determine all the information contained in a system that could then be sent via a classical channel from Alice to Bob.

The quantum channel is totally different. As mentioned, it uses a strange feature of quantum mechanics: entanglement. Einstein, together with two young colleagues, Boris Podolsky and Nathan Rosen, suggested in 1935 that two particles, two systems, may be connected in a very intimate way through quantum mechanics. This connection, or entanglement, is much stronger than any connection in any classical system, or any connection between any two everyday objects, for that matter. To explore what entanglement is, let's consider a little science-fiction story.

ENTANGLED QUANTUM DICE

It is sometime in the far future, at least the year 2100. A friend brought you, as a birthday present, the latest gift shop gadget. He hands you a little machine of shining blue. The plaque on it reads QUANTUM ENTAN-GLEMENT GENERATOR. It has a button on top (Figure 8). The instructions say that the machinery pops out a pair of dice if you hit the button.

So, you hit the button and you hear the two dice fall, each into its own little cup. You pull the cups away, and each has a lid on it, so you cannot see the dice. Your friend tells you that they are covered so you won't disturb their quantum state. He encourages you to look in. You open the first cup and you see the top face showing ⚃. You look in the other cup—also ⚃.

"What a fortuitous accident!" your friend says. "They show the same number."

But you are not impressed. You know that such events are not so rare. "Throw one die, and it shows, say, three. Then you know the other

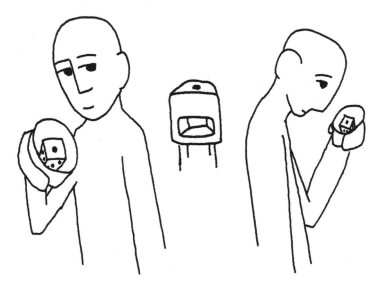

Figure 8. The science-fiction Quantum Entanglement Generator produces pairs of entangled dice (*top*). These dice do not show a number before they are observed. When one die is observed, it randomly chooses to show a number of dots. Then, the other distant die instantly shows the same number. The dice are quantum mechanically entangled, a phenomenon that Albert Einstein called "spooky action at a distance."

die will show three in one out of six cases, that is, with a probability of one-sixth." you say.

"Fair enough," your friend says. "Put the dice back into the machine."

So you put the dice back, return the cups, and hit the button again. Now, opening the first one, you get ⚅. The other die also shows ⚅. Your interest is awakened. You repeat the procedure. You get ⚀ and ⚀. Repeat. ⚁ and ⚁. And so on. After twenty attempts, you give up. They always match. So you start thinking that maybe the machine somehow loads the dice each time so that they are not fair dice anymore. Having gotten suspicious, you take the dice after the last throw, where they had shown ⚅ and ⚅. Then you throw the first die by hand. The die falls face ⚂ up. You throw the second die by hand—it shows ⚀. You keep throwing. They each show independently the usual random sequence of numbers. So they cannot be loaded.

Your friend has been watching you with a smile that is getting bigger and bigger. "You see, this is quantum entanglement at work. Each of the two dice is a fair die. It shows you randomly one of the six numbers. But if you put them both back in, the machine entangles them. When they come out of the machine, both dice always show the same number."

Your children have now become interested.

"Let's see what happens if I open the one cup over in the kitchen!" Your daughter runs away. She comes back, reporting that she got ⚂, and you open your cup. Also ⚂. Your children keep running all over the house and out into the backyard. No matter how far they run and where they look into the cup, their die always shows the same number as yours.

After some time the children get exhausted and you all get together to relax. Your friend explains, "Now you see why Albert Einstein called entanglement 'spooky' and he wanted to have a physics without it. He would be very surprised today to learn that our quantum computers use entanglement all the time."

Of course, such entangled quantum dice do not exist yet. But pairs of particles, like photons, electrons, protons, atoms, or even small atomic clouds have been shown to exhibit this strange feature of entanglement.

When we talk about such entanglement between particles, we have to ask ourselves which property is entangled. It turns out that many different properties of particles can be entangled. For photons, the particles of light, the easiest way to get entanglement is to entangle their polarization. We will learn more about polarization later. It suffices to say here that polarization is the way light oscillates. When a photon pair is entangled in polarization, this means first that before observation, neither of them is polarized in any way—just like the dice that do not carry any number on their faces before being observed. But when you observe one of the photons, it randomly assumes a specific polarization, say, horizontal or vertical, which means that the electric field oscillates horizontally or vertically. Then, in one kind of entanglement, the other photon must show exactly the same polarization when observed.

So the general rule is that some property of one of the two entangled particles must exactly correlate with a corresponding property of the other entangled particle. There are many possibilities. Another possibility is that the energies of the two particles are entangled. For example, the sum of the two energies of the two particles is constant, but neither particle carries any well-defined energy before being observed. When you measure the energy of one of them, it randomly assumes some specific value—and the other one then has the corresponding value to make up the constant sum, no matter how distant it is.

These are just two different examples of entanglement, in order to illustrate the idea.

THE ORIGINAL TELEPORTATION PROTOCOL

In the teleportation paper mentioned above, the six physicists suggest using a pair of entangled particles as the quantum channel (Figure 9). Needless to say, the proposal deals only with the teleportation of individual particles, and not of people. We again meet the two players, Alice and Bob. Alice has a particle in some quantum state that in general is unknown to her, and she wants Bob to receive a particle in exactly that state. We know already that measuring her particle and telling Bob the result does not work because the measurement changes the state of the original.

Let's now discuss the quantum teleportation protocol in some detail. At first sight the reader might find it somewhat complicated, but there

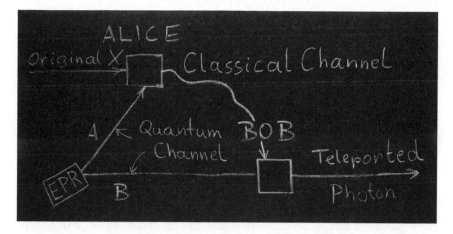

Figure 9. The principle of quantum teleportation. Alice teleports the quantum state of the original, X, over to Bob. She does this with the help of two communication channels—the quantum channel and the classical channel. The quantum channel consists of an auxiliary pair of entangled particles, A and B, which are created in the EPR (Einstein-Podolsky-Rosen) source. Alice gets the auxiliary entangled particle A, and Bob gets the auxiliary particle B. Alice now performs a Bell-state measurement, jointly on the original X and on her auxiliary particle A. A Bell-state measurement is a procedure in which A and X become entangled with each other. Through this entanglement, X loses its own private properties. These are transferred over to Bob's particle B. An interesting twist is that in the entangling Bell-state measurement of A and X, four possible results can occur randomly. Correspondingly, Bob's particle B will assume one of four different specific states. Each one already contains all the information about X, but this information cannot be deciphered by Bob yet. Alice therefore has to send Bob the information about which of the four results she obtained. This is done via the classical channel. Using this information, Bob can modify his particle B such that it ends up in exactly the same state as the original X. This concludes the quantum teleportation.

is relief. In the following chapters of the book, we will come back to the protocol again a number of times and discuss these things more. The original teleportation protocol goes as follows.

1. Alice and Bob anticipate that they might want to teleport a particle sometime in the future. They generate for themselves auxiliary pairs of entangled particles. Alice gets the A twin from each auxiliary pair, and Bob gets the B twin. Now, the important point is that the twin particles A and B are pairwise entangled, which means that if measured in the same way, the particles will show the same result: they will turn out to be identical.

This entanglement connection between the two twin particles is the "spooky action at a distance" that Einstein did not like. But entanglement works, no matter how far the two particles are separated from each other. This is the quantum channel.

2. The next step is that Alice receives a "new" particle, the original X, to be teleported. She now takes one of her A particles out of its box and performs a difficult task. She entangles X, the original particle, and her twin particle A with each other. We will have to come back later to the question of how this entanglement procedure works. Let's just accept for now that it can be done.

Alice's entangling measurement is called *Bell-state measurement*. (In honor of the Irish physicist John Bell, we call such entangled states *Bell states*.) So, what does this entangling do for us? It actually means that the original X loses its own private properties. There are various different kinds of entanglement. The simplest one, the one we're discussing here, is that the two particles X and A will turn out to be identical if they are measured. Neither the original X nor Alice's twin particle A have any features of their own left after they become entangled with each other.

This procedure of entangling the two particles is a hard thing to grasp because it is difficult, if not impossible, for humans to imagine objects that don't have their own private features and are identical nevertheless. But that is the essence of entanglement. None of the particles entangled with each other have their own private properties, but should they be observed, they will turn out to be the same. Yet the features they show then will not be features that were there before the measurement.

3. When Alice does her entangling procedure, what happens to Bob's twin particle B? With some luck, it turns into a particle that is identical to Alice's original X! This can be seen by a very simple line of logic. Originally, Alice's and Bob's twin particles were entangled. This means that they will turn out the same if they are observed, but they don't have their own private properties before they are observed. Then, through Alice's entangling procedure, her original particle X and her twin particle A may become identical. Thus, with B identical to A and A identical to X, we conclude B identical to X.

So, we have the situation that entanglement occurs twice in the procedure. Alice's original and her twin particle become entangled, and Alice's and Bob's respective twin particles are entangled similarly at the beginning. This is the reason why Bob's twin particle ends up with the properties of the original. All the features of the original particle have been teleported over to Bob, and his twin particle turns out to be just the same as Alice's original particle was. His twin becomes the teleported particle. Furthermore, Alice's original loses all its private features because it has become entangled.

We have to expand our considerations by one more important point. We have learned that Alice does a joint entangling procedure on the original X and her twin particle A. Through this Bell-state measurement, the two become entangled. But we have also learned that any measurement in quantum mechanics always means some randomness, some uncontrollable feature. When we looked at the electron, its place, the position where we found it was random. So, what is the random element here? It turns out that the specific way the two particles become entangled is random. There are actually different ways in which two particles can become entangled. Just think of our example of the pair of dice. The entanglement could be that the two dice are always the same, for example, ⚄⚄. But there are also other kinds of entanglement. One specific possibility could be that the numbers they show always add up to seven. In that case, possible results on the observation of the entangled states would be ⚀⚅ or ⚁⚄ or ⚂⚃. So these are two very different kinds of entanglement, and we might very well assume that our future quantum entanglement generator—our dice entangling machine—has a small switch where we can set which of the two kinds of entanglement we want the machine to pop out.

In the teleportation proposal, our six colleagues chose quantum mechanical two-state systems, systems that may have two possible states in a given experiment. For example, particles of light, photons, might have two possible colors, red or blue. In such a situation, it can be shown by quantum mechanics that four different kinds of entangled states exist. There is no point in going into the nitty-gritty details about why this is so. To arrive at a general understanding of the situation, suffice it to say we must accept this as a fact.

For the teleportation experiment, this means that Alice's entangling measurement procedure might result in four different entangled states for the teleportee particle X and her twin particle A. The important point now is that depending on which specific entanglement Alice obtains for her two particles X and A, Bob's twin particle B ends up in one of four different states. It happens that one out of four times, Bob's particle B is identical with the original right away. The other three times, Bob has to apply a small modification to his particle, depending on the result that Alice achieved. And this is why we need the classical channel. Alice uses the classical channel to inform Bob which of the four results she obtained, so that he can apply the proper modification to his particle to make sure that out pops the original.

It is important that Alice's entangling procedure for particles X and A does not give her any information whatsoever about the original state of X. It is essential that the quantum channel provides the possibility of transferring the properties of the original over to Bob. This can only be done—and here we now come back to the Heisenberg problem—without the features of the original being measured or determined in any way by this procedure. Therefore, there is no contradiction with the Heisenberg argument. Indeed, neither Alice nor Bob, after the teleportation procedure has been performed, knows what the state of the original X is or what the state of Bob's particle B is. And this is OK, since they don't need to know that information.

It is important to notice that Alice has no influence over which of the four entangled states appears in her Bell-state measurement procedure. So, it turns out that all four possibilities have the same probability of 25 percent. Each of the four possibilities turns up in one-quarter of all Bell-state measurements.

The way the procedure is tested in an experiment is to send a well-known state into Alice's apparatus. This can be done by a third party, say, Victor (Figures 10 and 11). Bob then checks whether he always gets a system in the state Victor claims. Victor could, without saying so beforehand, choose particles in various different states and then tell Bob which feature of the particle to measure. That way, they can confirm that the teleportation procedure actually succeeded.

The first experiments used the polarization of photons. By now, experiments have also been done with other properties. But before we go into these details, we should familiarize ourselves better with the concept of entanglement and the concept of polarization. In particular, the concept of entanglement and its counterintuitive features are quite mind-boggling and will keep us busy for a while.

That way, we will learn in more detail how teleportation works, and we will come back to the experimental verifications and the conceptual consequences later.

Figure 10. Alice and Bob want to perform a teleportation experiment. For that purpose, they produce (*top right*) an entangled pair of dice. This means that none of the faces of either die shows any dots. But if the dice were to be observed, they would show the same dots on their top faces. Victor initially has a die that shows on its top face ⬛. He passes it over to Alice, asking her to teleport its state over to Bob (*bottom*).

Figure 11. Alice then performs a Bell-state measurement on the two dice she has now, the one she received from Victor and the one from the original entangled pair. That way, these two now become entangled. Through this procedure, the faces of Bob's die obtain well-defined numbers of dots. In certain cases, these are actually the same as the original handed by Victor to Alice (*top*). Bob shows his die ⚀ to Victor, to prove to him that teleportation succeeded (*bottom*).

ALICE AND BOB IN THE QUANTUM LAB

The corridor, which had seemed to stretch for miles in front of the two undergraduate students when they rounded the corner, is suddenly at an end. Alice nervously twirls a lock of her blond hair before she extricates her hand from the tangles and resolutely knocks on the door in front of them. On the friendly "Come in!" from inside, Bob opens the door and they step into the office of Professor Quantinger. Desk and coffee table are cluttered with physics books, copies of articles, and pieces of experimental equipment. Professor Quantinger looks up from his computer screen, and his preoccupied frown turns into a smile. "Alice, Bob, what can I do for you? Ready for the exams on Wednesday?"

"Well . . . yes, but . . ." Alice stutters and flushes. "We'd like to—well, our first-year course, Physics 101, is very exciting, but there's so little quantum. We would like to know more about . . ." Bob comes to the rescue: "Is there anything else we can read on quantum physics apart from the reading list? Could you recommend a few books?" Professor Quantinger's friendly smile broadens into a grin. "I think I have a better idea. How would you guys like to work with a real quantum experiment? There's nothing better to help you understand quantum physics than some hands-on experience."

Alice and Bob exchange a glance. "You mean, like in lab class?" Alice whispers. Lab class can be quite frustrating—sitting in front of apparatus that looks too complicated to comprehend and just collecting some data in order to write up a report and that's it. There is often not enough time to really understand what's going on there.

The professor seems to understand. "No, this is a real scientific experiment, a graduate project that was published recently. John, my graduate student, set it up. He is presently writing up his Ph.D. thesis, so he is still around to help you. He'll do all the setting up of the equipment, aligning it and making sure it works as precisely as it should."

Now, that's a different story for Alice and Bob—but frightening in another way. With the little knowledge they gained in Physics 101, will they be able to meet the challenge? Resolve straightens Bob's shoulders. After all—what a compliment! Alice's eyes light up at the prospect of doing something *real*. Two young voices simultaneously exclaim, "Awesome!"

Professor Quantinger, sensing the uneasiness behind the resolve to do well, sits back, and his voice becomes soothing. "In science, you always start from observation. We want to find out how nature is. We are curious to get at its inner workings, so we want to know what it is that keeps everything going, keeps it from falling apart and keeps it running. But to do that, we first have to find out how it operates. We have to watch the phenomena. We have to watch what it is doing. That's the same in all of science.

"If you don't first observe what is really going on, you might make grave mistakes. Human imagination is a great thing. For a long time, people had all kinds of fantastic ideas about what kinds of animals might exist on unknown continents, like in Africa. If you look at some of the old pictures, they are quite fantastic.

"The beautiful part is that today, though we have had many more chances to observe and learn exactly what the animals in Africa look like—and to find out that not all of those creatures people dreamt of exist—we can still see that the world is very rich and that, actually, in many respects nature is richer than man can contemplate. Consider, for example, the thousands of different species of orchids that exist all over the world . . ."

Alice and Bob start getting uneasy in their chairs. The professor, realizing this, apologizes. "Sorry, I got carried away. You have to know that in my spare time, I like to study how rich nature is and collect specimens of unusual plants from all over the world. So, let's get back to our physics. I want you to study the phenomena first and make up your own story of what might be going on."

"But we have no knowledge of quantum physics at all," Bob says hesitatingly.

The professor continues: "I feel it is actually great that you guys have no deep knowledge of quantum physics. So you will find your way all on your own. It won't be very complicated. I'll give you an opportunity to study a very simple experiment and figure out how it might work. The experiment consists of three pieces of apparatus." The professor goes to the blackboard and makes a few sketches (Figure 12).

"We call this piece of apparatus the source. It sends something, some 'stuff,' through special cables to two pieces of equipment at two labs at opposite ends of the campus. There, the detectors will register something whenever the stuff—I don't want to tell you yet what this stuff is—arrives."

"But how can we ever find out what this stuff is really made of and how it behaves?" Alice asks. "Can we look inside the source, what it does, and inside the detectors?"

"Well," the professor replies, "you could actually look inside the source and the detectors, but you won't see much, just some pieces of equipment. The stuff that flies around is too tiny to be seen."

"So our job is hopeless," Bob says.

"Not really!" The professor smiles. "There is a lot you can do to find out what goes on. On each of the detector stations, you can control some switches and you can see what the detector tells you when it registers. Alice, you might control the detector on the river side, and Bob, you control the detector on the town side. John will make sure every morning that everything works perfectly, and you guys just collect data and try to find out what it is. So, come back Monday morning, and John will introduce you to the labs."

The following Monday, they meet John at his office and he tells them, "I have already set up the source, and I have checked that both detectors are working properly. I can do this from my computer here in my room. You know, I set up the whole experiment two years ago. I used it for my Ph.D. dissertation, testing quantum nonlocality over large distances."

At the mention of quantum nonlocality, Alice's and Bob's faces beam. They had already heard about Einstein's "spooky action at a distance." They had heard mysterious buzzwords like "the Einstein-Podolsky-Rosen paradox" and "Bell's inequalities," and they had heard some senior students say that these are some of the most interesting top-

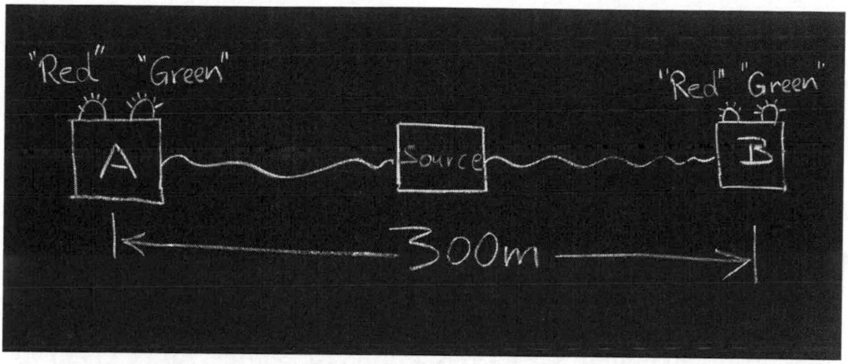

Figure 12. Arrangement of the experiment done by the students Alice and Bob. The source sends some "stuff" to two measurement stations, station A in Alice's laboratory and station B in Bob's laboratory. The two measurement stations are separated by a distance of about 300 meters. At each of the measurement apparatuses are two lamps, a red one and a green one, which light up according to the result of the measurement performed by a detector inside each box.

ics a person can work on in modern physics. So apparently, the professor wanted them to work on exactly the kind of questions they were really interested in.

"So, let's go see the pieces of equipment!" John suggests.

Alice says, "The source, I understand, is just in the basement of this building?"

John smiles at that suggestion: "You'll be surprised!" When they arrive in the basement, they see only a large black box sitting on the table with cables going in and out, and a computer next to it.

"What's inside?" Bob asks.

"Well," John says, "the professor told me you guys should not know too much about the equipment. You should find out for yourselves what is going on. So I put this black box around the source. All I can tell you is that there is some sort of laser inside, and the stuff that is produced enters two glass-fiber cables." He points at them. "The computer controls the whole setup, and the glass-fiber cables run away from it, this one over to your lab, Alice, and this one over to Bob's."

"What do we have to do at the computer, some adjustments?" Alice asks.

"No, that's not your job. I'll make sure that everything runs. You don't have to worry about that at all."

"How boring!" Bob says.

Anyway, he and Alice are starting to get curious. "What a strange experiment this is!" Alice says. "We don't get any clue to what's going on, and we have to find out ourselves what comes out of the black box? How can we do that? It's a hopeless job!"

"That's exactly what science is all about," John mentions. "You don't know what's going on in the beginning. You have to make careful observations, play with the little equipment you have, make up your own story, and try to find out whether it's right."

"But scientists can often look in detail at what they investigate," Bob says.

"Well, not always!" John replies. "Just think of the astronomers. All they have is some light or radiation coming from distant stars, and some piece of equipment they can manipulate here on Earth. And look at what fascinating information astronomers have been able to find out about the universe."

"Oh, I see!" Alice answers. "That's what we have to do. We have our pieces of equipment that we can play with, and we have to find out what comes over from the source."

"Exactly!" John grins. "That's exactly the job. I was rather uncertain in the beginning, when the professor told me about the experiment, but now I feel that you two have a pretty good chance of figuring out what's going on. So, we'll now walk over to your two labs, and I'll tell you how to operate the equipment there."

Outside it has started to rain a little, so Alice asks if the operation of their equipment might depend on the weather, as the professor told them that something is being sent from the source to the two detectors. Perhaps this "stuff" might not arrive there if it is raining or snowing heavily.

John smiles. "The source and the two detectors are connected via underground glass-fiber cables, so that everything is completely independent of any weather. What might happen is that changes in temperature might influence the experiment somehow. Actually, I had to study that in my dissertation, just to make sure that you don't see any unwanted effects, and I did not see any."

Arriving at Alice's lab, they are surprised at how simple it looks. Sitting on a table are just a computer and very few pieces of equipment: some lenses, glass fibers, mirrors. A few cables run from the computer to the equipment. That is all. John says, "You can see the cable coming in here from the wall. This is the glass-fiber cable that connects to the source. At the other end, we have a similar setup. As you also see, the computer is connected to the Internet, so it can talk to the computer controlling the source and to the computer at Bob's side. This was necessary for my experiment, but it is also important for yours here, since the three computers talk to one another every few minutes to make sure that all pieces of equipment are working properly."

"And what happens," asks Alice, "if they don't work properly?"

"We don't have to figure out what it means and fix it when it isn't working properly, do we?" guesses Bob.

"Correct," John says. "You don't have to worry about what might go wrong. You'll learn to build your own experiments and fix them later in the course of your studies. Actually, in case the equipment does not work properly, the computers can automatically make some small adjustments to set it up again. There can be situations where something goes completely awry and cannot be corrected by the computers. In that case, you might see a message on the computer screen or you yourself might get a hunch that something went wrong. Just call me if that happens. I'm around most of the time, writing my thesis up. I am sure I can fix it. Bob, you know where your lab is? It's just the same. Here's the key. Have fun, guys." He starts to leave.

"So, that's all?" Alice exclaims. "What are we supposed to do now? The professor told us that we can observe some results and that we can operate some switches!"

Turning back, John replies, "Oh, I forgot to explain how to operate the equipment. You have a switch here. A similar one is over at Bob's side. This switch has three positions—plus, zero, and minus $(+, 0, -)$. Bob has the same switch over there. *Plus*, *zero*, and *minus* are just the names we give to the positions of the switch. By setting the switch in a certain position, you can determine what it is that the detector is actually observing. These are three different properties of the incoming stuff, and again, you don't have to worry what it is. You can also see here that the computer screen indicates which of the three positions you have chosen. There are also these two lamps—red and green—which light up to indicate the result observed."

Alice asks, "There are actually only two results possible?"

John explains, "Yes. For each of the positions—plus, zero, and minus—the incoming stuff can register at one of two detectors. Whenever an incoming particle is registered by a detector, it sends out an electric pulse that you can actually hear." He turns on a small loudspeaker, which makes "click . . . click . click . . . . click . . click . click . . . . click" sounds.

"Each of those clicks corresponds to a particle," John tells them. "One detector is connected to the red lamp, and the other one is connected to the green lamp. The lamps were actually set up for you guys so you can visualize the results more easily. For the experiment itself, the detectors are also connected to the computer. The computer remembers both the setting of the switch—plus, zero, or minus—and which of the two detectors registers, red or green, for each time there's a click. You can also see all this on the computer screen.

"The time when a click is registered is measured by a very precise atomic clock. Don't worry. You don't have to know how an atomic clock works. This atomic clock runs all the time. It's connected to the computer, and it tells you exactly at which time a particle was registered in the red or green detector. So you see, at present, the switch is set at zero and the red and green lamps are flashing in a rather random fashion. All I want you to do is to find out what is going on here. OK, you guys are on your own now. Bob, you know the way to your lab?"

"Sure!" answers Bob, and off he goes.

ALICE AND BOB'S EXPERIMENT—THE FIRST STEPS

After John leaves and Bob has walked over to his townside laboratory, Alice starts to play with her apparatus. There is not very much she can play with: the switch with the three positions—plus, zero, and minus—and the lamps that register the individual events, red and green. There is also the computer that remembers the time when each event happened, which color it was, and which switch setting was chosen.

Furthermore, she discovers she can also simply count how often a particle is registered for the given setting of the switch, in both the green and the red channels separately, and put that on the display. She also has the option of printing all this out in a neat format. Watching the apparatus, she notices that the flickering of the red and green lamps looks quite irregular, but on average, there is about one flash every second, either green or red, in some random sequence. So she decides to find out whether the red and the green lamps flash equally often or whether there is a difference between them. She decides to set the apparatus to count the events in both the green channel and the red channel for 200 seconds. The result she gets is red 105, green 98.

"Interesting," she thinks. "There are more counts in the red. Maybe there is a slight difference in the detectors—they like to register red more often than green." So she decides to count again for 200 seconds. This time she finds red 101, green 106. "The opposite. This time, there are more green events than red ones," she thinks. Repeating that procedure a couple of times, she finds that there seems to be no preference whatsoever. On average, each of the two detectors, the red one and the green one, register about 100 events in a 200-second set. Sometimes there are slightly more red ones; sometimes there are slightly more green ones. The whole result looks pretty uninteresting.

"Oh," she thinks, "maybe I was unlucky in having worked with the zero setting in the beginning. I should try the other ones, plus and minus."

So again, she sets the counting time to 200 seconds and observes how often the red and the green lamps flash this time. They appear equally irregular, as before. And this irregularity is actually confirmed by counting. Both the red and the green lamp each flash on average about 100 times in the 200-second interval. And again, for both the plus and minus settings, sometimes there are slightly more green flashes; sometimes there are slightly more red flashes; occasionally they are equal.

So she concludes that the whole experiment is really totally boring. The two lamps flicker along, together giving on average about one count per second, but apparently, there is no structure at all. Furthermore, setting the switch to any one of the settings does not seem to influence whether the red or the green lamp flashes more often. Actually, setting the switch to any of its three positions—plus, zero, and minus— does not seem to have any consequences. She concludes that something must be wrong with the experiment. Apparently, the computer has gone wrong and setting a switch does not change anything. Also, the flickering of the lamps seems to be completely random, not telling any story. Anyway, it is time to go to her advanced class, Physics 102, where she is already a little late. Today Professor Quantinger's lecture is on the polarization of light.

THE POLARIZATION OF LIGHT:

A LECTURE BY PROFESSOR QUANTINGER

"You have already learned about the frequency and the wavelength of light," Professor Quantinger begins his lecture. "Frequency and wavelength are directly related to each other, and they determine the color of light. In addition, there is also another property called polarization, which we will now study experimentally. Let's look at a light source, say, a simple incandescent lamp. Here I have two polarizers—two sheets of Polaroid—a special kind of plastic" (Figure 13).

The students, who were chatting at the beginning of the lecture, immediately become quiet when they realize that the professor is about to demonstrate an experiment.

"If we look through one of these polarizers"—and Professor Quantinger holds one of the polarizers in front of his eyes—"we notice that there is somewhat less light passing through. So some of the light is absorbed by the polarizer, but a gray sheet of plastic would do the same. So, let's put these two polarizers together, one behind the other, and look at the light coming through. The amount of light coming through now depends on the relative orientation. If the second polarizer is rotated with respect to the first one, the intensity varies from zero, with nothing coming through, at one orientation, to a lot coming through at another orientation, a right angle away. Actually, these polarizers have some marks that indicate how they were fabricated, and if these marks coincide, the two polarizers are said to be oriented parallel with each other. As we can see, all the intensity that comes through the first one also passes through the second one."

The professor points at the screen where the students can see a spot of light. This spot is the light that passed through the two polarizers. Its

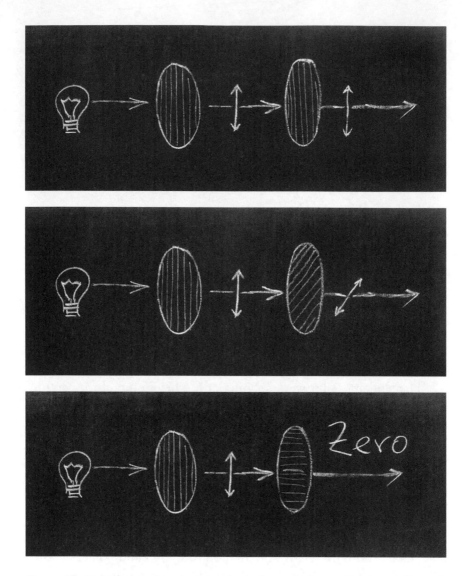

Figure 13. Light from a lamp passes through a polarizer, which lets only vertically polarized light pass through. Behind the polarizer, the polarization oscillates up and down along the direction shown by the double arrows. If a second polarizer is arranged parallel to the first one (*top*), all the light that passed the first polarizer passes through it also. If the second polarizer is oriented at right angles to the first one, nothing passes through (*bottom*). Finally, if the second polarizer is at some arbitrary angle, part of the light passes through (*center*), but at a reduced intensity. Most important, the light coming out has lost all information about its initial polarization, and it is always polarized according to the orientation of the last polarizer it passed through.

brightness changes according to the orientation of the second polarizer, which the professor rotates around.

"The explanation is simply that light carries polarization. As we learned earlier, light can be understood as oscillating electric and magnetic fields. The electric field is actually a concept invented to describe how electric charges move around. Basically, if you have two like charges, say, both + or both −, they repel each other. Different charges attract each other. How does one charge act from a distance on the other one? The explanation is that one charge creates an electric field, which the other charge then experiences. Depending on whether this other charge is positive or negative, it is pulled along the electric field or pushed away.

"For simplicity's sake, we'll look only at the electric field. After the first polarizer, the electric field of the light coming out shows oscillation, but the electric field oscillates up and down along one direction only. This direction is determined by the orientation of the polarizer.

"In fact, the light coming from the lamp has all kinds of polarizations, but the polarizer only lets electric fields pass that are oriented along one direction. So, if the second polarizer is inserted and is oriented parallel to the first one, all the light that came through the first one can now pass through the second one. If we now rotate the second polarizer by 90 degrees, the light coming from the first one cannot pass through at all anymore, because it oscillates in the wrong way.

"It is actually interesting how such polarizers work. They are built up of long chains of molecules with parallel orientation. An electric charge in the plastic can move easily along these chains of molecules but has a very hard time moving in a direction at right angles to them. So what happens if light comes in? It tries to bring the electric charges into an oscillatory motion, because the electric field itself oscillates. It goes up and down. If the electric field oscillates parallel to the orientation of the molecules, the electric charges can move easily. This motion takes up a lot of energy from the light. In the end, the light loses so much energy that it is completely absorbed by the plastic. In contrast, if the electric field oscillates at right angles to the chains of molecules, the electric charges cannot really move much. This reduced motion takes only very little energy from the light so the light beam hardly gets reduced and passes through.

"If we now put our second polarizer at some oblique orientation with respect to the first one, part of the light passes through. If we measure carefully, we find out that exactly half of the energy of the light comes through when the second polarizer is oriented at 45 degrees with respect to the first one. An angle of 45 degrees is just halfway between zero and 90 degrees, between parallel and right angles. But the naïve expectation that the energy coming through is proportional to the angle is not correct."

Here, the professor raises his voice. "If the naïve expectation were correct, then at an angle of 22.5 degrees, which is just half of 45 degrees, we would have three-quarters of the intensity coming through. But it turns out that as much as 92 percent passes through. Three-quarters of the intensity passes through at an angle of 30 degrees, and at an angle of 60 degrees, only one-quarter. It turns out that the amount of energy coming through as a function of relative angle is actually the square of the cosine of this angle. This is called Malus's law, after the French physicist Étienne-Louis Malus, who lived from 1775 to 1812 and discovered this law." With these words, the professor draws a cosine curve on the blackboard, which represents Malus's law (Figure 14).

"The light coming originally from the incandescent lamp is unpolarized. It is a mixture of all possible polarizations, which means that all directions of oscillation of the electric field occur.

"If we pick out any one such direction, the electric field can be divided into a component that is parallel and one that is at right angles to the polarizer axis. The component parallel to the axis is absorbed, and the component at right angles passes through.

"Let's pick out one direction of polarization, for example, light polarized at 45 degrees." The professor draws a figure on the blackboard (Figure 15). "What will happen with that light if it meets, say, a vertically oriented polarizer? Then, the electric field can be split into two components, one parallel to the polarizer axis and one at right angles that is, orthogonal, to it. The parallel component passes through the polarizer while the orthogonal component is absorbed. That way, in the case shown in the figure, exactly half of the energy passes through.

"So far, we have discussed polarizers where only one polarization passes through and the other one is absorbed in the material that makes up the polarizer. In some experiments, we want to have both polarizations available, and this can be done using what is called a polarizing beam splitter (PBS). The PBS looks like a cube, but it actually consists of two wedges glued together. These wedges are not ordinary pieces of glass; they

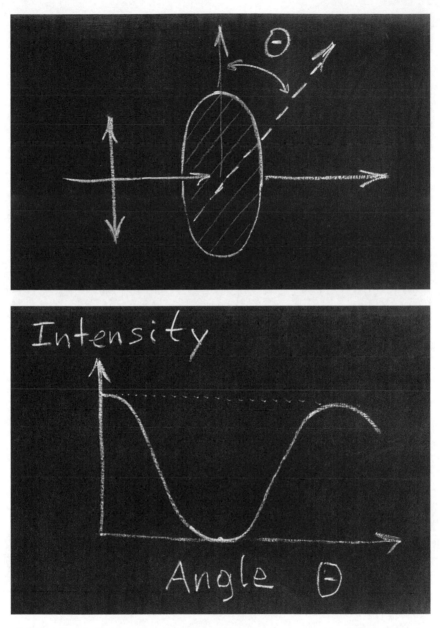

Figure 14. Vertically polarized light impinges on a polarizer rotated by the angle theta (Θ) (*top*). If the angle theta is zero degrees, then all light passes through. If the angle is slowly increased, less and less light comes through until, at 90 degrees, nothing passes through anymore. Increasing the angle further, the intensity increases again. This is shown in the bottom curve, which represents Malus's law.

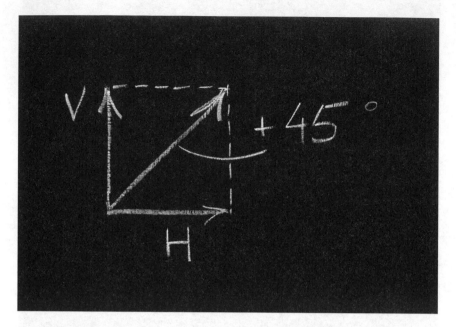

Figure 15. Light polarized at 45 degrees can be decomposed into two equally strong components, one horizontally (H) and the other one vertically (V) polarized. The light polarized at 45 degrees is a superposition of these two components. Suppose that light now meets a polarizer vertically oriented. Then, the vertical component passes through and the horizontal component is absorbed by the polarizer.

are made of a special kind of crystal where the two polarizations of the light have different velocities. It turns out, as can be seen in the figure" — the professor again draws a figure on the blackboard (Figure 16) — "that one polarization passes through and the other one is reflected to the side. That way, we can have all the intensity of an incoming light beam available in two beams polarized at right angles to each other.

"The polarizations we have discussed so far are cases where an electric field oscillates up and down along a direction. Therefore, this is called linear polarization, and the orientation of the line along which the field oscillates characterizes the specific polarization. So, for example, one could have horizontally (H) or vertically (V) polarized light with horizontal and vertical being the orientation of the direction of the oscillation. Every polarization can be split up into a part that is horizontally polarized and a part that is vertically polarized, their relative fractions depending on the orientation.

"Likewise, we could take two other directions that are orthogonal to each other. For example, we could take a new horizontal axis rotated by 30 degrees from the horizontal and a new vertical axis rotated by 30 degrees from the vertical. Then, we could split up any polarization along these new angles, in general with different portions than before. People often refer to the orientation of the axes as frame of reference, so we might speak of the frame of reference oriented at zero degrees with respect to horizontal or of the frame of reference oriented at 30 degrees. There are an infinite number of possible orientations of such reference axes.

"To conclude the story about the polarization of light in terms of electric fields, there is not only linear polarization, but there are other, more complicated forms. The most important one is circular polarization, where the electric field actually does not oscillate up and down, but goes around in a spiral way. Evidently" — and the professor tries to show big spirals with his arms — "one can have right-handed and left-handed circular polarization."

THE POLARIZATION OF INDIVIDUAL QUANTA OF LIGHT

"So far, we have learned to understand the polarization of classical light," the professor continues. "What you really have to remember is that you can build up any arbitrary polarization as a superposition of

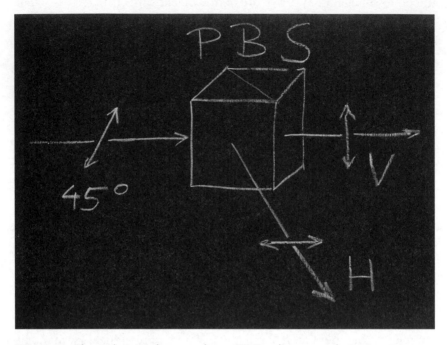

Figure 16. The polarizing beam splitter (PBS) splits any polarization into two components, a vertical one and a horizontal one. In the figure, this is shown for the special case of 45-degree polarization, which is split up into equal parts of horizontal and vertical polarization.

others," he says, pointing to the blackboard again (Figure 15). "But now, the story becomes really interesting." And the professor again raises his voice. "Let's now consider that light really consists of individual particles, of quanta of light, also called photons."

After these words, the lecture hall becomes absolutely silent. For the first time the students are learning something about quantum physics in their introductory class. They all know that this is something interesting, and they are surprised that the professor starts now to introduce quantum physics. They had always thought that they would have to know a lot of mathematics to understand anything in that field.

"If we look again at our light beams, they can be understood as consisting of many quanta of light, of many photons. So then some of our findings can be easily understood. If the two polarizers are parallel, all the photons that pass through the first one also pass through the second one. This is clear if you realize that the photons in between are polarized along the direction given by the first polarizer."

The professor points at another figure on the blackboard (Figure 13), which he had left there. "Equally well, we can understand the bottom part of the figure. If the polarizers are oriented at right angles, all the photons that pass through the first one are absorbed by the second one.

"So far, understanding the quantum world seems to be very simple." The professor pauses and gives the audience a broad smile.

"But the really quantum mechanically new phenomenon comes in when we consider the case where the two polarizers are oriented neither parallel nor at right angles to each other, but at some oblique angle, say, at 45 degrees." He points at the middle of the figure (Figure 13).

"There, we have learned that half of the energy passes through, and we understood this in terms of decomposing the light fields into a part parallel and a part orthogonal to the polarizer orientation.

"But let us now consider what this means for the photon concept. The important message is that an individual photon is indivisible. So each photon has to either pass through or be absorbed by the polarizer. We learned earlier that half of the light passes through. This means that just half the number of photons that passed through the first polarizer also pass through the second one. So far, there seems no conceptual issue in understanding what goes on even at the quantum level.

"But let's now consider an individual particle of light, an individual photon, after it has left the first polarizer. We know that the photon is polarized along a direction given by the orientation of that polarizer. What will happen to the photon then? Will it pass through the second polarizer, or will it be absorbed? Evidently, if the second polarizer is oriented parallel to the first, the photon will pass through. If the second polarizer is oriented at right angles, the photon will be absorbed. But what will the photon do if the polarizer is oriented at some oblique angle— for example at 45 degrees, just halfway between the parallel and right angle positions [Figure 13]? What will happen to the photon then? Clearly, our old considerations don't apply here. When we had an intensive beam, half the energy passed through and half was absorbed by the polarizer for that orientation. But now, we have a single photon, a single quantum, and quanta cannot be divided into parts. So, it is impossible for half a photon to pass through and half a photon to get absorbed [Figure 17].

"The only two possibilities are that the photon either goes through or is absorbed. It turns out that the photon has a fifty-fifty chance of doing either. So the probability of passing through is 50 percent. If each photon has a 50 percent chance, then if you take many photons, just half of them will go through, which means that half the energy passes through and the other half is absorbed, just what we found earlier for a light beam.

"We now ask a most important question. This is actually one of the most important questions we can ask in quantum physics." The professor paces up and down very rapidly. "What determines whether a specific photon passes through or is absorbed? Does the individual photon meeting the polarizer know whether it will pass through or be absorbed, or more broadly, how does it know what to do?"

The professor stops his pacing in the middle of the lecture room in front of the blackboard and faces the audience, exclaiming, "It turns out that there is no rule whatsoever for that. There is no explanation of why an individual photon does this or that. There are no hidden properties of photons that would explain what they do, no tiny marks on a photon or whatever. This probability is fundamental and not further explainable in any way.

"In quantum physics, we can only give explanations for large numbers of particles and how they behave, for what we call 'ensembles.' Only in very rare cases can we tell beforehand what an individual particle will do. In general, there is no way to explain what an individual

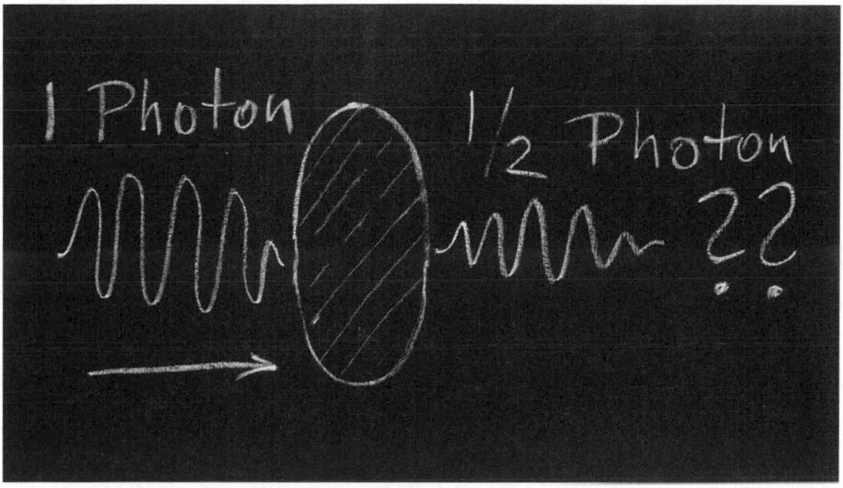

Figure 17. A single, vertically polarized photon impinges on a polarizer oriented at 45 degrees. Can this mean that half a photon passes through? According to quantum physics, this is not possible, since photons are light quanta, and thus indivisible.

photon does, and we have good reasons to believe that this is not just our ignorance, but that this fundamental role of probability is a basic feature of how the universe works.

"To me," the professor continues, "this is one of the most important discoveries ever made in physics. Just imagine what physics, or science in general, does. We have tried for centuries to look deeper and deeper into finding causes and explanations, and suddenly, when we go to the very depths, to the behavior of individual particles of individual quanta, we find that this search for a cause comes to an end. There is no cause. In my eyes, this fundamental indeterminateness of the universe has not really been integrated into our worldview yet."

The professor pauses and looks around. In some students' faces he clearly can see his own excitement. For others, he is not so sure if they understand the importance of his message.

"Anyway," he continues, "let's go back to discussing the physics of the behavior of individual polarized photons. Let's go one step further, and consider a stream of photons polarized at 45 degrees encountering a two-channel polarizer [Figure 18]. Any individual photon will be transmitted by the polarizing beam splitter or reflected off to the side with equal probability. Therefore, each photon has a 50 percent chance of passing through the polarizer and a 50 percent chance of being reflected.

"Now, a most interesting point concerns the polarization of the photons after the polarizing beam splitter. All the photons that have passed through will be polarized vertically, and all the photons that are reflected to the side will be polarized horizontally. So, each individual photon ending up, for example, in the straight-through beam will have forgotten whether it was initially polarized at 45 degrees or whether it was already horizontally polarized before it hit the polarizing beam splitter.

"We now come to an important question," the professor continues, pointing again at the figure on the blackboard (Figure 18).

"We just found that half of the photons pass through and the other half get reflected at the polarizing beam splitter. But when does an individual photon decide which path to take? The way I drew the figure seems to suggest that the photon makes this decision when it is inside the polarizing beam splitter. If that were so, it would at that moment decide whether to continue traveling straight on or whether to take the

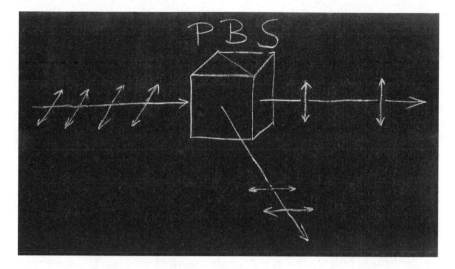

Figure 18. A stream of photons polarized at 45 degrees strike a polarizing beam splitter (PBS). Half the photons are transmitted; the other half are reflected. The transmitted ones will all be vertically polarized, and the reflected ones will all be horizontally polarized. Each photon is indivisible, and it will be found in either of the outcoming beams. Yet, whether a specific photon is reflected or transmitted is completely random.

deflected path out of the beam splitter. Such an explanation would be easy, wouldn't it?" And the professor looks around.

"Sure! What else could happen?" a student in the back row who was following intently says, more to himself.

The professor, hearing this, smiles again and says, "Well, that explanation would be too simple. If it were that way, we could understand quantum mechanics more easily."

The professor apparently enjoys the attention that follows his words, and he writes on the blackboard the word "superposition."

"Now, you are about to learn the most important concept of quantum physics," he continues, "quantum superposition. A photon actually does *not* decide at the moment when it leaves the beam splitter what to do. In contrast, after it has passed through the polarizing beam splitter, it is in a superposition of both possibilities. In a sense, though such wording is very dangerous, the photon is in both beams at the same time. These two superposed possibilities are whether the photon has passed through or has been reflected. Think of a wave. The incoming wave is split into two waves, one passing through and one reflected.

"But this wave we are talking about is a very abstract wave. It's not a real wave that is found somewhere in space. It's a wave whose only purpose is to determine with what probability a detector would register the photon if we put such a detector into the respective beam. People talk of a probability wave.

"So, let's suppose we put photon detectors, one each, into both beams. Such detectors would register a photon. They would make a 'click' if a photon is present in the beam. Clearly, one of the two detectors will register the photon. Which one, though, is completely open. There is a probability of 50 percent for each one to register the photon." The professor hits the blackboard with his hand in excitement: "And only in *that* moment does the photon decide which path it took. Until that very moment, all that can be said about the situation is that there is a superposition of both possibilities.

"The American physicist John Archibald Wheeler expressed this once in a very strange way. He said, 'The photon takes both paths, but it takes only one path.' That is a very provocative way of expressing the situation," adds the professor, smiling. "But it meets the core of the problem." He faces the audience: "And if you are now confused, welcome to the club!"

"Let's recapitulate," the professor continues after a pause. "Whenever we send a single photon polarized in the way indicated in the figure into a polarizing beam splitter, the photon emerges in a superposition of two possibilities, transmitted or reflected. But interestingly this is also a superposition of two polarizations. Before, the photon was polarized at 45 degrees. Now it is either polarized vertically, namely, in the case when it is found in the transmitted beam, or it is polarized horizontally, namely, in the case when it is found in the reflected beam. Therefore, if we put one detector each into each of the emerging beams, the photon will be registered in one of them. That photon will therefore also have a definitive polarization, vertical (V) or horizontal (H), depending on which beam it is found in.

"But most important, if one detector has registered the photon, the other one will certainly not register the photon, because we have only one photon and not two."

"How do we know that a photon makes only one click?" a student asks.

"Very good question!" the professor replies.

"Theory tells us that, but more important, experiment completely confirms it. People have done such experiments with individual photons, and they see that this is exactly what happens. Only one of the two detectors registers a photon, and never both. That is actually the final confirmation of the quantum nature of light.

"But we might ask: How does the second detector know that it should *not* register the photon? Because until the photon is registered, there is an equal probability for both detectors to detect the photon. In the language we just learned, the photon was in a superposition of both possibilities. The way we explain this in physics is that we say, 'The superposition collapses in the moment when the photon is registered by either one of the detectors.' Superposition breaks down all over space. Einstein imagined a more complicated situation where the photon is actually spread out over many more possibilities than just the two beams, and if we put a detector anywhere, and we register the photon, the superposition collapses instantly all over space.

"This is instantaneous," the professor emphasizes. "It is faster than the speed of light. Imagine one detector up there at some distant star"— and he points with his hands up to the sky—"and the other detector over there at some other distant star. Suppose this detector registers the photon. The superposition immediately collapses. The other detector cannot now register the photon. It instantly knows that it is no longer allowed to click.

"This collapse of the superposition is another quantum phenomenon Einstein did not want. He specifically did not like the fact that this was something faster than the speed of light. But we have to accept that both detectors never fire together, because the photon is indivisible. This kind of experiment was first performed in the laboratory by the American physicist John Clauser in 1974 and again in an improved way by the French physicists Philippe Grangier and Alain Aspect in 1986. But it has not been performed over large distances yet. It might be interesting to do this in the future.

"The problem does not arise if, like most physicists today, we regard quantum physics only as giving us probabilities but refuse to consider any realistic picture of a wave spreading out. But Einstein didn't like that position. He wanted physics always to describe the physical reality out there, and not just give us probabilities. Once he wrote in a letter to Max Born that he was convinced that God 'does not play dice.' Well, I believe"—the professor smiles—"that the Lord actually loves to play dice.

"God seems to have taken liberties in creating the universe such that he does not know what will happen in certain cases, for instance, in the individual quantum event just discussed. God took the liberty of making the world more interesting for himself. But that's a different story," he says, smiling.

"You might now ask, How do we know that the photon is in a superposition of these two possibilities? Do we have some experimental proof for that? Why isn't it that the photon decides when it is inside the beam splitter which path to take? Actually, so far I have not given you any proof. We will do that now.

"Let's consider a slightly more complicated situation," he says, making a new drawing on the blackboard (Figure 19).

"So, we have here two polarizing beam splitters," the professor explains, "and at the first one, on top, as we have already learned, something interesting happens. The photon emerges in the superposition of being in both paths, and in two polarizations, vertical (V) and horizontal (H). Then, we put these two beams back together at another polarizing beam splitter. What will happen?

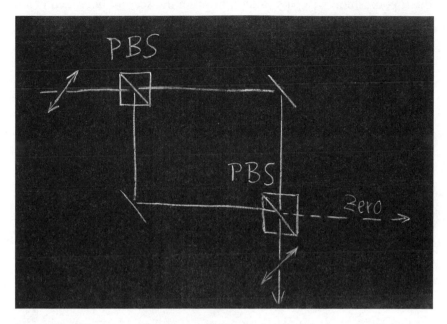

Figure 19. At top, a single photon polarized in some direction, say, 45 degrees, meets a polarizing beam splitter (PBS). The two emerging paths are then redirected by mirrors to meet again at some point, where we place another polarizing beam splitter (PBS). Of the two emerging beams, one is empty—there is no photon at all—and in the other beam, we again have a photon polarized at 45 degrees. This is the result of quantum superposition.

"Now, let's follow in detail through what happens. Along the upper path, the photon is vertically polarized. Therefore, that vertical part of the superposition will pass straight through the second polarized beam splitter. Remember [Figure 16], our polarizing beam splitters work such that it is the vertical component that is transmitted.

"Now, let's consider the lower beam path, the path that the photon takes when it is reflected. There, the polarization is horizontal, and therefore, that horizontal component will be also reflected by the second polarizing beam splitter.

"So, in the end, both components of the superposition will end up together in the beam emerging downward, and nothing will end up in the beam emerging to the right from the second polarizing beam splitter. So, we have in this downward beam a vertical and a horizontal component that are put together.

"What does this mean? Evidently, we reassembled the original polarization at 45 degrees, because the polarization at 45 degrees is just a superposition of the horizontal and the vertical" (Figure 16).

"Why does this little experiment, which actually has been performed in many laboratories by now, prove that the photon is in a superposition after the first polarizing beam splitter?" the professor asks. But he is now so deeply in his thoughts that he does not make contact with the students anymore. He just keeps talking along.

"Suppose this were not the case. Suppose," he says, pointing again at the earlier figure (Figure 18), "that each photon were to decide immediately after the first polarizing beam splitter whether to take the straight-through or the deflected beam path and therefore be vertically or horizontally polarized. Well, if that were the case, the individual photons would arrive as vertically or horizontally polarized at the second polarizing beam splitter. They would still end up in the beam going down, but there would be no reason for them to build up a superposition when passing through the second polarizing beam splitter. So this outgoing beam would just be a mixture of the horizontal and the vertical, half of the photons being horizontally polarized and half of them being vertically polarized. This is very different from a beam where each photon is polarized at 45 degrees. It is only by superposition, the fact that somehow each photon takes both paths, that we arrive at each photon being polarized at 45 degrees in that outgoing beam. This is a definitive proof of superposition for an individual quantum particle, and not just a classical wave." The professor raises his voice at the last words and makes a long pause.

"What I told you today," the professor continues after a short glimpse at the large clock in the lecture hall, "are some of the most important facts of quantum physics. Of course, you have to learn a lot of mathematics to see fully what is going on. But the two concepts that you learned, *probability* and *superposition*, will always accompany you whenever you do quantum physics, whether as an experimentalist or a theorist.

"To emphasize again, quantum physics gives us only probabilities for future events. There is no explanation whatsoever for why a specific result occurs in a specific measurement situation. Quantum physics does not explain, as in our case, why a specific photon is detected in that beam and not in another one behind the first polarizing beam splitter.

"But quantum physics does make very precise predictions for probabilities. In our case this means that if we have many photons coming in, we know that half of them will pass through and half of them will be deflected. So quantum mechanics makes very precise predictions for ensembles.

"There are actually situations where quantum physics also makes precise predictions for individual events to happen. You have one example here in front of your eyes." He points again to the last picture on the blackboard (Figure 19).

"Quantum physics predicts with certainty that each photon will end up in this downward beam right behind the second beam splitter, and no photon will end up in the beam emerging to the right. This is the only case where quantum mechanics makes definitive predictions, namely, in those situations where the probability for an event is either one or zero, where the event will happen either with certainty or never.

"The other important concept we learned is that of quantum superposition. In general, the situation is described in a complicated way where a particle is in a superposition of many different possibilities that might occur in the experiment. In our case, there are just two possibilities, namely, the two beams after the first polarizing beam splitter. The measurement is then one of the two possible results—more generally, one of the many possible measurements results. Furthermore, when this happens, the quantum mechanical state collapses, and none of the other possibilities can any longer occur.

"Finally"—and the professor raises his voice because some murmuring has already started in the lecture hall, as he has reached the end of the class period—"I want to make you a little more curious. The kinds of things we learned today were investigated by people just out of philosophical curiosity. They wanted to know whether nature is really as strange as quantum physics describes. Most interestingly, and to the surprise of everyone, the early experiments not only confirm that individual quantum particles behave in such a strange way, but they also lay the foundations for new technology. Today, we talk about new concepts of computing and communication, where the quantum behavior of individual quantum particles is essential. You might already have heard about the quantum computer, which can be immensely faster than all existing computers, about quantum cryptography, by which one can send messages in an absolutely secure way, and about quantum teleportation, by which you can send a quantum state of a system over large distances."

"Beam me up, Scotty!" a student in the first row pipes up.

"Well," the professor says, smiling, "it's not quite that, but actually even more interesting. But you have to wait a little to learn all this in detail."

The students applaud. Some come up front to try to get some more information out of Professor Quantinger about quantum cryptography, quantum computers, and quantum teleportation, and slowly, the lecture hall empties.

ALICE AND BOB DISCOVER TWINS . . .

After class, Alice meets Bob and tells him of her impression that her apparatus was malfunctioning. There was no way for her to influence how often the red and green lights flickered by changing the settings of her switch. "Maybe the switch has no influence at all on the whole setup, or it's broken," she suggests.

Bob has a similar impression. Whatever he did with his switch did not change what the detector lamps told him. He had also tried to make sure that what he was measuring really had something to do with the stuff coming from the source. So he had unplugged the cable coming from the source to his equipment, and lo and behold, the two lights did not register anything anymore.

"So," he concludes, "the flickering must have something to do with the incoming stuff, because at least it stops when I disconnect the cable. But apparently, the apparatus does not measure any interesting property of what comes in." So they decide to visit John and tell him that something is missing or wrong with the setup.

They go up to John's office, and Bob starts to explain how he had noted precisely how many times the red or the green lamp flashed, but that the flashing did not seem to have any structure at all. It appeared as if the red and green lights would light up in a completely disorderly way, sometimes this one and sometimes that one. When he then switched to the other positions, he would see the same thing again—no order whatsoever. Furthermore, nothing was registered when he disconnected the cable.

Alice adds her story, telling how she had counted the number of flashing lamps over 200 seconds and how, on average, each of the lamps showed registration of a particle about 100 times during that period independently of the switch setting.

Alice and Bob expected John to show some compassion and get worried, but to their surprise, they notice that very soon after the beginning of Bob's explanation, John starts to smile! The smile gets broader the longer they talk.

John is a very patient listener, so he lets them explain everything, asks questions about exactly what they had done, and then says: "Congratulations! Very good work. Everything you told me is quite correct. I am impressed."

Alice answers unbelievingly, "So the apparatus is not broken? These unstructured data are what we are supposed to observe?"

John replies, "This is exactly what I expect the network to do when it's working properly."

"But then, there is nothing to be measured!" Bob exclaims.

"No, there is," John says. "And with you guys working so properly and asking the right questions of the setup, I am sure you will find it. Just be patient."

"No hint?" Alice pleads.

"Well, maybe I should help you a little," John answers. "There is one feature of the experiment you did not make use of yet. It is the fact that your measurement statistics are both connected to the same source. But the rest you should find out for yourselves. You will be much more excited then. I have to go off to teach a recitation class. Give me a call anytime you want to talk about your next results." And off he goes.

Alice and Bob are completely perplexed. What should they do? They have found out—and their results are confirmed by John—that there is no structure at all in what happens at either station. And to make the situation worse, apparently this is the right way for the particles, or whatever it was that came from the source, to behave.

"If this random clicking is what the quantum is all about, it is much more boring than I had expected!" Alice exclaims.

"Maybe we should leave it and get a project that is more down to earth," Bob says. Mumbling to himself, Bob tries to figure out what the fact that both stations are connected to the same source might imply. Maybe they should look at the relationship between the data they observed at both stations. But what kind of correlation could there be? They had to find out by playing with the equipment again.

The next morning, both Alice and Bob are a little late, as they have become less enthusiastic about the whole project. But nevertheless, they show up around half past nine and start to make their observations. After some playing around, Alice calls Bob and tells him that her setup is still behaving the same way it did the day before. Bob tells her that the situation is the same for his experiment. They decide not to worry, because they were told by John that this is exactly how the apparatus should work.

"Well," Bob says, "there is this one small hint John gave us, that we didn't make use of the fact that both our stations are connected to the same source. But what would that do for us? What can we do?"

"I have an idea!" Alice exclaims. "Let's just compare at what times our detectors register something. Maybe there is some relationship. We should be able to measure very precisely the times when one of the lamps flickers, because we have this display on the computer that tells us at which time each click happens, and each of our setups has a very precise clock, down to a nanosecond, which is a thousandth of a millionth of a second. So let's just register a set of such events," Alice suggests on the phone.

But Bob says, "With so many clicks to observe, we have to make sure we start at the same time." Bob realizes that the computers also allow them to start the counting procedure simultaneously, and Alice discovers that they can set the computers for a fixed period of time.

So that is what they do. They both set their computers to start together with the time arbitrarily set at zero, and they both set the apparatus to count for 100 seconds. On the basis of what they saw the day before, they expect about 100 events, roughly half of them red, the other half green. After 100 seconds, they have their data on the computer screen, which looks quite messy. It is a long list of entries, one below the other.

Each entry corresponds to a single event when one of the detectors registered something. The entry consists first of the time, then a record of the switch setting (+, 0, or −), and then a record of the lamp that would light, red or green. This is written down for each event. So a typical entry in that table looks like this:

$$22.327033758 + G$$

This means that 22.327033758 seconds after the start of the counting, the green detector registered something while the switch was set at plus

(+). So Alice calls Bob and tells him, "Let's compare our results. First, let's see if we get some particles at the same time."

Bob is ready for a coffee break, and he suggests printing the two lists out and meeting at the cafeteria.

Alice, being a little more cautious, says, "Maybe something went wrong with the measurement we just did. Just to be on the safe side, let's make a few more."

So they both collect data three more times. Each time, both pieces of apparatus are set at the identical starting time. In the end, they obtain four lists, each for 100 seconds. Then, they go off to get their lattes.

"Here we go!" Alice says and puts her four lists on the table.

"Our typical ambitious undergraduates!" the waitress, a graduate student herself, says, smiling. "You guys were born with the right genes to do science."

Bob replies, "Well, that's just what we're going to find out now. So far, it does not look too good, as we don't understand the quantum."

The waitress comforts them. "I remember reading about this famous professor—what was his name? I forget—who said, 'I think I can safely say that nobody understands quantum mechanics.'"

"Yes, that was Richard Feynman," Alice explains, "and he had actually received the Nobel Prize for some variant of quantum mechanics, which, as our professor mentioned, is a very successful one in calculating all kinds of experimental results for elementary particles."

"But I guess we won't get the Nobel Prize for not understanding our experiment," Bob says. "Anyway, let's look at our lists." As the lists look quite messy, Alice and Bob decide first to check whether they have the same number of entries within each second. In the first batch of data, it turns out that for red and green together, Alice has 102 and Bob has 98.

"Oops," Bob says. "My apparatus is worse than yours. Let's look at the others."

In the second set of data, Alice has 95 and Bob has 100. It turns out that none of the pairs of data sets have exactly the same numbers of registration events; sometimes Bob has a few more, sometimes Alice. Interestingly, the average is the typical 100 Alice had found the day before by just playing with her apparatus alone.

"But if we measure the same kind of phenomena, we should get the same number of events," Bob says. "So something is wrong with our apparatus."

Alice, having become a little more cautious after incorrectly assuming the day before that her apparatus was broken, suggests, "Well, I am sure if we go back to John and tell him that the apparatus is wrong, he'll say that this weird behavior is just what he expects. And he'll again be proud of having set his apparatus so that it behaves in such a weird manner."

So they sit there for a while and stare at the lists. "Well," Bob suggests, "maybe we should look at the exact times when we both registered something."

"But we can't compare them, because the total number of events registered is never the same, so some must be missing," Alice says. But having no other ideas, they look at the details of the data in their first list (Table 1).

TABLE 1

Alice's list		Bob's List		
00.382234518	- G	00.882031592	0	R
01.129527532	- R	02.240987810	0	G
02.240987809	- R	03.097710128	0	G
03.300187990	- G	. . .		
. . .				

The two entries in the first line are different, as are the two entries in the second line, and so on.

So this finding—that there is no apparent correlation even in the times—does not add to their enjoyment. Bob's mind starts to wander, and he absentmindedly stares at the paper in front of him, when something suddenly catches his eye. Alice's time entry in the third line is nearly identical to his entry in the second line! There is a discrepancy in just the last digit, that is, in the nanoseconds; otherwise, the entries are the same. So they get very excited and circle the two entries (Table 2).

"Here the source must have emitted two particles together, which then arrived at the detectors at the same time!" Alice exclaims.

TABLE 2

Alice's list		Bob's list		
00.382234518	-G	00.882031592	0	R
01.129527532	- R	(02.240987810	0	G)
(02.240987809	- R)	03.097710128	0	G
03.300187990	- G	. . .		
. . .				

They get excited, so they look for more such identical or nearly identical registrations at the same time, and they find many. In the end, it turns out that for about one-fifth of the entries on each of the lists, there is a corresponding entry on the other list that has exactly the same time up to a few nanoseconds.

Alice suggests that this nanosecond difference might just be an intrinsic imprecision of the apparatus. "After all, every clock, if you use it for very precise measurements, must become uncertain somewhere, and evidently, for our setup, that's in the nanosecond range. So let's not worry about that. So, what should we do with the data? We now know that about a fifth of the events on the two sides are correlated with each other. They happen at exactly the same time."

Bob suggests, "This must be because the source emits two particles at exactly the same time. They must be like twin brothers."

"No, twin sisters," says Alice.

"Never mind—like twins," Bob says, smiling, "and then they move through these glass fibers, as John explained, to our two measurement stations and arrive here at the same time because they are registered at the same time."

Alice says, "But then the glass fibers must be of exactly the same length, which is very unlikely."

Bob gets on his cell phone and calls John. "John, are the glass fibers from the source to our labs of equal length?"

John exclaims, "Congratulations! You found something very important. Obviously, you are already at the essence of the business, looking at correlations between both of your labs. Yes, I took great care, for your sake, to have them be of equal length, even though the two labs are not at precisely the same distance from the source. I coiled up a few feet of glass-fiber cable on Alice's side, because her lab is a little closer to the source. The two fibers are of equal length within an inch or two. I guess you guys found coincidences."

Bob says, "Oh, that's what you call it when two events happen at the same time at these two distant locations?"

John replies, "Exactly. Did you find out anything else?"

Bob answers, "Well, we found out that this does not happen all the time. Only in about a fifth of the cases do we have coincidences. In the others, an event on my side does not have a twin brother on the other side."

"And," Alice yells into Bob's cell phone, "an event on my side does not seem to have a twin sister on the other side."

John's voice betrays a smile. "Gee, I am impressed. You guys are really doing precise work. I expect coincidences exactly 22 percent of the time."

"How can you expect that?" Bob asks.

John answers mysteriously, "Well, you have to make up your own story about this, trying out a few explanations on your own. I won't explain. Keep going. Bye, now." And he hangs up.

Alice and Bob start to think about what John might have meant by expecting to find coincidences only 22 percent of the time. They end up with two alternative explanations. One explanation is that the source sometimes emits single particles, and sometimes twin particles together at the same time, and that it emits twins just 22 percent of the time. The other possibility is that the source always emits particles in pairs, but that some particles get lost or the detectors are not good enough to register all the particles, and it turns out that 22 percent of the coincidences is what they are finally left with. Alice and Bob realize that there is no way to know for sure which of the two happens.

"But," Alice suggests, "if I look only at my data, not at yours, all the events look pretty much the same. There is a random flickering, and the particles arrive at more or less the same rate, so there is no reason to assume that the source emits two different kinds of stuff."

Bob is excited. "That's a very good idea! So we can guess that the source emits pairs all the time, and we only register both particles roughly one-fifth of the time."

They immediately call John again and tell him the idea. "I'm quite impressed!" he says. "You just discovered the collection efficiency loophole problem of all these experiments."

"The *what*?" Alice asks.

"Well," John answers, "maybe I already told you too much, but there is indeed a problem in the sense that while the source really emits particles as pairs all the time, we are able to collect the twin of a particle that has been registered in only 22 percent of the cases."

"And why is that a problem?" Alice asks.

"Well, that's where I would be going too far in telling you more than you should know about the experiment." And he hangs up again.

Excited as they are, Alice and Bob again scrutinize the data closely to see if they can find out more.

"Maybe we should pick out only those cases where we have coincidences and look just at them," Alice suggests.

So they mark all their coincidences, twenty in all. They now decide to pair the outcomes, to compare whether the red or the green lamp flashed. As expected, because of their random nature, each one flashed about half of the time, and this was true for both Alice's list and Bob's. More precisely, Alice has eleven red and nine green on her list, and Bob has ten red and ten green on his list.

Alice says, "Let's now look at which color goes with which. Maybe we have some cases where we get the same color and some cases where we get different colors."

So, looking at the colors, they find out that the situation isn't very exciting. It turns out that sometimes they both have the same color, and sometimes they have different ones. Checking all the lists more precisely, they find the combination red-red eight times and the combination green-green nine times. It appears that identical results on both sides are more likely than different ones. There is only one red-green — that is, Alice got red and Bob got green — and green-red appears only twice.

"That seems to tell us something. But what?" Alice asks. "Let's recapitulate what we have so far. In some cases, namely about 22 percent, we register twins. The other cases we can ignore for the moment. And we register all four combinations of colors. Like colors seem to appear much more often than different ones."

Bob says, "And looking at the tables, there does not seem to be any pattern of how they follow each other. How can we find out more? Could there be more structure in the system?" Bob scratches his head, and Alice stares into space, twirling a lock of her hair. So, as it is now late in the morning, they decide to call it quits for the day and get back to the apparatus the next day to see what they can do.

During the night, Alice does not sleep very well. She keeps dreaming of the flickering red and green lights and of the computer screen and the printed-out data they had just analyzed. In her dream, she has a very vivid picture of the printouts. She cannot exactly see the data printed, but she sees the pieces of paper in front of her with the list of entries. Vaguely, she realizes that there are three columns. There is one for the time, and one for the result, red or green, but there is a third one they had not paid attention to. She remembers that John had told her there was also the switch setting, which was registered on the printout. In the dream, this third column fades in and out and changes all the time, from + to − to 0 and so on. Evidently, there is something important she cannot quite grasp yet. But in the dream, she also does not worry about this, because she and Bob had found out that their results do not depend at all on the switch setting when they looked at the individual results, but never mind.

The next morning when she wakes up, she remembers her dream. She calls Bob immediately on the phone and tells him her idea that the switch setting might have some relevance.

"I looked at my printouts," she says, "and they all have the switch set at minus. What about your switch?"

Bob looks at his. They all had the switch set at zero. These were the settings their switches just happened to have when Bob and Alice arrived at the lab the day before. On that day, they did not care about playing with the switch setting.

So now, the plan is obvious: to get back to the lab to see if the switch setting has any influence on the correlations of both numbers. Maybe it influences the times at which the equipment registers pairs. Maybe there is also a way in which the results depend on whether they use the same switch settings or not.

So off they go, after a quick breakfast, to their individual labs. But what to do, how to play? Each apparatus has three different switch settings available, +, 0, and −. Altogether that makes for 3 × 3 = 9 possible combinations. Quite a mess. So they decide to make it a little easier and start first with those cases where they set their switches to the same setting, starting with plus-plus. Again, they count for 100 seconds, printing out all the results as before, including the time and color of the output. Again, they do this for four periods, each of 100 seconds' duration.

This time, they are so eager that they do not want to move, but call each other on the phone. Knowing from the day before what is important, they quickly identify those data where both of them have registered a particle. The result is the same as before. In only about one-fifth to one-fourth of the cases they both observe an event. In the other cases, there is an event either on Bob's side or on Alice's side, but no twins. So they each decide to scratch out on their lists those cases where they do not have twin events and compare the colors for those left.

"For the first result," Alice says, "I have green."

Bob says, "I have green, too."

The second one: Alice, "red"; Bob, "red."

The third one: Alice, "red"; Bob, "red."

The fourth: Alice, "green"; Bob, "green." They get very excited. Obviously, they always obtain the same color result on both sides.

Comparing all four lists, they see that there is only one case where they do not have the same color. In that case, Alice has red and Bob has green. In all others, their colors are exactly the same. Alice and Bob are sure that they have discovered something important. They decide to discard this one event as some spurious measurement error. They had both learned in their laboratory courses that there is always some measurement error in any experiment. The professors tell them again and again that errors are effectively unavoidable. Thus they conclude that if both settings are on plus, the results on both sides are equal. Further, they conclude that the combinations red-red and green-green occur equally often.

For Alice and Bob, it is an attractive idea to assume that the source sends pairs of particles out that are somehow produced in an identical way. These pairs would be such that they either both register green or both register red. But it is still unclear to them what role the switch position really plays. One thing they had learned is that they get the same result on both sides if both switches are on plus.

The next step is to try the same with both switches on zero to see what kinds of coincidences appear then. They find the same pattern. Whenever both sides register a particle, the results on both sides are identical, either red-red or green-green. The same results are obtained when they put both switches at the minus position.

. . . AND INVENT HIDDEN PROPERTIES

"Let's try to find an explanation for what we observed here," Alice proposes. "Obviously, the two particles are identical. They were born with the same properties."

"But what kind of properties? We don't have any idea," Bob says.

But Alice does not worry about this. "It could be any kind of feature. We don't have to know, as long as it is something that determines which of the two detectors registers, and therefore which of the two lamps lights up, the green one or the red one."

"Maybe you're right," Bob says. "We could simply assume for now that both particles carry some sort of instructions with them, and when your particle meets the detector, it looks at the instructions to see whether it has to go to the red or the green detector, and my particle does the same."

"That's a great idea!" replies Alice. "We now know that the instructions on both sides have to be equal for the same setting of the switches. This is just like identical twins."

"Why?" Bob wants to know.

"Quite simply," Alice says excitedly, "because for identical twins, we know why they are identical. They carry exactly the same genes, and these genes determine the color of their hair, the color of their eyes, and many other features."

Now Bob picks up on her excitement. "And maybe these instructions we're talking about are just like identical genes for the two particles. Each particle must then carry a gene for the measurement result if the switch is on plus, another for the measurement result when the switch is on zero, and a third for the measurement result when the switch is on minus. If these genes are the same, both particles carry the same instructions and we get our coincidences."

But Alice replies, "Strictly speaking, the individual particles don't need to carry instructions or genes for each of the three positions of the switch because they are measured for only one of these three features."

"You're right," Bob says. "The source only has to know in which of the three positions the respective switch is and send a particle on its way with the instructions for that position."

Alice looks out the window, and then sounds a note of caution. "This might not be enough, because we could change the switch quickly, at the last moment, when the particle is already on its way. Then the particle would not know what to do, because it would have been born with the instructions for the wrong switch setting. If that's so, then the two particles must carry instructions for all three switch settings."

Bob considers that idea for a while and says, "Well, we can't do that experiment because we cannot be fast enough given the high speed of light. But maybe this is technically possible. So let's for the moment accept your assumption and see what John says to our ideas."

So they call John, who immediately hears that they are totally fascinated. When they tell him about their observation of perfect correlations, he is really impressed. And he is even more impressed when they tell him about their idea of hidden instructions that the particles carry.

"It's really great that you not only found the perfect correlations, but that you also discovered the reasoning of Einstein, Podolsky, and Rosen."

"What do you mean by Einstein, Podolsky, and Rosen?" Alice wants to know.

"Well, I don't have the time now to explain it to you, but drop in at my office later this afternoon so I can tell you what it's all about."

JOHN'S INTRODUCTION OF EINSTEIN, PODOLSKY, AND ROSEN

Alice and Bob settle down comfortably in their chairs in John's office, and he starts his small lecture.

"Earlier, we learned that Einstein received his Nobel Prize for a very important contribution to quantum physics, the explanation of the photoelectric effect, and not for his relativity theory. We also discussed some of the criticisms raised by Einstein, particularly his criticism of the new role of randomness in quantum physics signified by his famous saying about God not playing dice.

"But in 1935 came Einstein's 'bolt out of the blue,' as his fellow physicist Léon Rosenfeld liked to call it. Einstein had been one of the leading physicists in Berlin. In the first third of the twentieth century, Berlin was one of the world's centers in science, the humanities, and the arts. Many scientists and artists who are famous today did their work there. Among them were the physicist Erwin Schrödinger, the chemist Fritz Haber, and the physiologist and medical scientist Otto Heinrich Warburg (who all won the Nobel Prize) and the architect Walter Gropius, to name just a few. All that ended when the Nazis seized power, which marked an end for freedom and led to the brutal persecution of many intellectuals, particularly those of Jewish descent.

"In March 1933, during a trip through the United States, Einstein decided not to return to Germany. He gave up his position at the Prussian Academy of Sciences in protest against the Nazi regime. After brief stays in Belgium and England, in October 1933 he traveled to America and remained there. In Princeton, at the Institute for Advanced Study, he continued to worry about the foundations of quantum physics.

"In 1935, together with the young physicists Boris Podolsky and Nathan Rosen, Einstein wrote a paper entitled, 'Can Quantum-Mechanical Description of Physical Reality Be Considered Complete?' This is a very philosophical-sounding title. Nevertheless, it appeared in volume 47, issue 10, of the *Physical Review*, a physics journal published by the American Physical Society."

John walks over to his computer and logs in to the homepage of the *Physical Review* (http://prola.aps.org) to get a download of the article (Figure 20).

"The paper itself is rather technical, and apparently, Einstein himself did not like it very much. Witness a letter he wrote to Schrödinger, where he says that for language reasons, this paper was written by Podolsky after many discussions. It did not turn out as good as he wanted it to be, but rather, the main point was buried, so to speak, under erudition."

Alice interrupts: "So, couldn't Einstein have presented his ideas in a more simple way, without so many equations?"

"He actually did," John replies. "In 1949 he wrote his *Autobiographical Notes*, and there he explains the essential points in a very simple and clear way.

"But let me try to explain the basics of the Einstein-Podolsky-Rosen paper to you. The paper is often referred to as the EPR paper, as scientists often use the first letters of the authors' names for papers that are important. By the way, it turns out that this paper is one of those whose importance increased steadily with time. A measure of the relevance of a paper is how often it is mentioned by others when they write their own papers."

"I already saw that in my undergraduate research studies. People in general quote others when they write their scientific papers. Why do they do that?" Bob asks.

"There are various reasons," John replies. "One is that it's good practice to mention other people's work if you use it in your own paper, to make sure that you don't give the impression of the work being yours. Another reason for quoting other papers is to give more credibility to your own position if it happens to agree with the position of another scientist. Also, you often want to show where you expect your own contribution to stand with respect to the work of others, particularly if you believe that you made a breakthrough other scientists have been searching for.

of lanthanum is 7/2, hence the nuclear magnetic moment as determined by this analysis is 2.5 nuclear magnetons. This is in fair agreement with the value 2.8 nuclear magnetons determined from La III hyperfine structures by the writer and N. S. Grace.[9]

[9] M. F. Crawford and N. S. Grace, Phys. Rev. 47, 536 (1935).

This investigation was carried out under the supervision of Professor G. Breit, and I wish to thank him for the invaluable advice and assistance so freely given. I also take this opportunity to acknowledge the award of a Fellowship by the Royal Society of Canada, and to thank the University of Wisconsin and the Department of Physics for the privilege of working here.

MAY 15, 1935   PHYSICAL REVIEW   VOLUME 47

Can Quantum-Mechanical Description of Physical Reality Be Considered Complete?

A. Einstein, B. Podolsky and N. Rosen, *Institute for Advanced Study, Princeton, New Jersey*
(Received March 25, 1935)

In a complete theory there is an element corresponding to each element of reality. A sufficient condition for the reality of a physical quantity is the possibility of predicting it with certainty, without disturbing the system. In quantum mechanics in the case of two physical quantities described by non-commuting operators, the knowledge of one precludes the knowledge of the other. Then either (1) the description of reality given by the wave function in quantum mechanics is not complete or (2) these two quantities cannot have simultaneous reality. Consideration of the problem of making predictions concerning a system on the basis of measurements made on another system that had previously interacted with it leads to the result that if (1) is false then (2) is also false. One is thus led to conclude that the description of reality as given by a wave function is not complete.

1.

ANY serious consideration of a physical theory must take into account the distinction between the objective reality, which is independent of any theory, and the physical concepts with which the theory operates. These concepts are intended to correspond with the objective reality, and by means of these concepts we picture this reality to ourselves.

In attempting to judge the success of a physical theory, we may ask ourselves two questions: (1) "Is the theory correct?" and (2) "Is the description given by the theory complete?" It is only in the case in which positive answers may be given to both of these questions, that the concepts of the theory may be said to be satisfactory. The correctness of the theory is judged by the degree of agreement between the conclusions of the theory and human experience. This experience, which alone enables us to make inferences about reality, in physics takes the form of experiment and measurement. It is the second question that we wish to consider here, as applied to quantum mechanics.

Whatever the meaning assigned to the term *complete*, the following requirement for a complete theory seems to be a necessary one: *every element of the physical reality must have a counterpart in the physical theory*. We shall call this the condition of completeness. The second question is thus easily answered, as soon as we are able to decide what are the elements of the physical reality.

The elements of the physical reality cannot be determined by *a priori* philosophical considerations, but must be found by an appeal to results of experiments and measurements. A comprehensive definition of reality is, however, unnecessary for our purpose. We shall be satisfied with the following criterion, which we regard as reasonable. *If, without in any way disturbing a system, we can predict with certainty (i.e., with probability equal to unity) the value of a physical quantity, then there exists an element of physical reality corresponding to this physical quantity*. It seems to us that this criterion, while far from exhausting all possible ways of recognizing a physical reality, at least provides us with one

"The idea is that the more often a paper is quoted, the more important its contents are. Certainly, this can be misleading, because if a paper contains a famous mistake, it might also be quoted very often. In the case of the EPR paper, the citations of the paper grew steadily over the years. Here, I created a graph of the number of citations made to the paper in articles received in the scientific literature [Figure 21]. As you see, in the beginning, after its publication in 1935, there were virtually none, and now, we are well over a hundred a year. This is very unusual. Most papers receive maybe one or two citations in all, and also, usually the number of citations declines as time passes."

Alice remarks, "That is really strange. How could it have been that such an important paper was nearly ignored in the beginning?"

John nods approvingly and continues, "The ideas of the EPR paper are very important for investigations done today in some of the most advanced physics research worldwide. But before I discuss the content of the EPR paper, maybe I should tell you a little about the broader context."

John leans back comfortably in his chair and raises his eyes up to the ceiling. "In physics, we describe phenomena by theories that we physicists construct. A theory in science is a very precise concept, much different from the colloquial meaning of the word 'theory.' In daily life, by 'theory' we often mean something like a hunch or a feeling about how something might work. A theory in physics is more precise. It makes exact predictions for possible future observations and for measurement results. And physicists are very immodest. They want to find theories that are more and more general, that is, theories that explain more and more. This approach has actually been quite successful in the history not only of physics, but of science altogether. To take an example not from physics, but from biology, Charles Darwin's theory of evolution has a very broad range of validity, all the way from the smallest viruses and bacteria to us humans.

"But, as I said, physicists are very immodest. They want to find a theory that, in the end, is able to explain everything. So the big dream is that there will be no physical phenomenon that cannot, at least in principle, be described by that final, universal theory of physics.

"One can certainly ask the question of whether such an attempt to find a Theory of Everything might not be too immodest, whether it is reasonable to expect to find such a theory someday or not. Some phys-

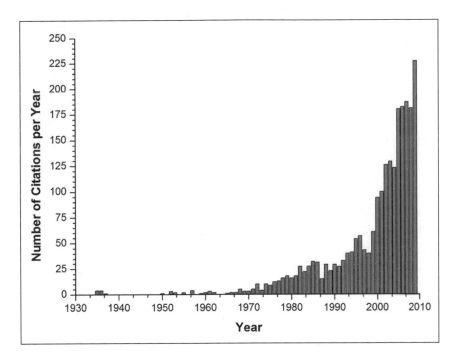

Figure 21. Number of citations to the EPR paper in articles received by the *Physical Review* every year. In the beginning, the paper was essentially ignored. Today, it has high impact.

icists believe it may be found very soon because physics and science in general have been able to build more and more theories capable of explaining many things that were outside the possibility of scientific explanation before. So these physicists argue that there is no reason why this extremely successful effort might stop someday.

"Still others argue that such a theory is still very far away, reasoning that the whole endeavor of modern science is really only a few hundred years old. Still other physicists argue that such a Theory of Everything might never be achieved for fundamental philosophical reasons, one of the reasons being that a physical theory always has to describe what is being observed, so it cannot include the observers themselves.

"So, these physicists argue, in order to describe ourselves we would have to be able to look at ourselves from the outside, which of course is impossible. Therefore, a Theory of Everything is impossible.

"But let's leave that discussion aside. It would carry us into very deep philosophical waters, which are interesting, but beyond the scope of our task," John continues.

"Independently of any such fundamental considerations, it is certainly legitimate to ask whether the physical theories that mankind has at a given time are a complete description of reality within the field for which they claim to be applicable. This is exactly the question the EPR paper asks about quantum mechanics."

"How would we find out whether a theory is complete or not?" Alice asks.

"EPR suggested that we have to look at physical reality. Einstein was a realist. For him, a fundamental tenet was that there exists a reality 'out there,' independent of us and independent of whether we observe it or not.

"EPR then suggested that in order for a theory to be complete, '*every element of the physical reality must have a counterpart in the physical theory.*'" John points to the quotation on the first page of the EPR paper (Figure 20).

"That's very heavy!" Bob interrupts. "What are these elements of physical reality? How do I know whether they're some physical reality or some other reality or whatever?"

"That's really the hard part of the job," John replies. "My understanding is that what EPR means here is simply objects around us in the world—anything that exists and that you can talk about. Elements of physical reality, for example, are objects like this blackboard, a person, and so on.

"Einstein, Podolsky, and Rosen themselves worried about this notion a lot, and they suggest that elements of physical reality cannot be found by pure thinking, but rather, they 'must be found by measurements.' So in our example, the physical reality of this blackboard can be based on the fact that we can describe its features, we can perform measurements of it, and we can measure its color and many other features. So, in general, EPR asks, When do we have an element of reality in front of us? How do we know that? It's certainly very difficult to give a complete definition of reality, and the EPR paper doesn't even want to attempt that. Rather, it gives a famous criterion for the existence of an element of reality: a sufficient criterion, not a necessary criterion."

"This always confuses me," Alice interrupts. "What do you mean by 'sufficient criterion'? And what do you mean by 'necessary criterion'? None of the professors in our lectures ever explained to us what these are, but they all use these words."

John answers, "Fair enough. Basically, a criterion allows you to recognize something. For example, how can you tell whether you have an elephant in front of you?"

And John points to a drawing of an elephant (Figure 22) he had done on canvas.

Alice is amazed. "I didn't know you were that good at drawing."

John answers, a little embarrassed, "Well, I was always interested in art, and I nearly became an artist, but somehow, my scientific inclination won out in the end. But let's talk about this elephant. Well, it has to have four legs, a trunk, huge ears, and two tusks. It also has to be gray and quite large. If the animal in front of you has all these features, you know for sure it's an elephant."

"But it does not *need* to have these features," Bob says. "For example, there are lots of elephants that have no tusks, because they have been removed."

"Or," Alice adds, "they can be baby elephants, so they don't have to be huge."

"Perfect!" John says. "What I listed for you before was a *sufficient criterion*. That is, if this animal has a trunk and four legs, is huge and gray, and has tusks and huge ears, you know for sure it's an elephant. But it's not *necessary* for the animal to have all these features. You could have

Figure 22. To illustrate the concepts of sufficient and necessary criteria, John points to various features of an elephant that could be used to establish the fact that the elephant is an element of reality.

an elephant that had lost a leg or was a baby or whatever. So, none of these criteria by themselves are necessary, but all together they are sufficient for you to make sure that you have an elephant in front of you."

"Now I *am* curious," Bob says. "What is the EPR criterion for an element of reality?"

"Hold on," John replies. "I'll have to read it to you from the original paper."

THE REALITY CRITERION

John picks up the printout of the first page of the EPR paper again and reads out loud (Figure 20):

"*If, without in any way disturbing a system, we can predict with certainty (i.e., with probability equal to unity) the value of a physical quantity, then there exists an element of physical reality corresponding to this physical quantity.*"

"Here, as in the earlier quote, even the italics are from the original. They considered their definition extremely important. Actually, it sounds a bit technical, and this is maybe what Einstein meant in his letter to Schrödinger about it being buried by erudition," John says. "The criterion sounds pedantic and stuffy, but in principle, it is simple. So, let's analyze in simple words what it says and what it does not say.

"First, what does it mean to 'predict with certainty'? We have to be cautious here and avoid possible confusion. Prediction for a physicist does not necessarily mean the same as for a prophet. It doesn't mean foretelling the future. It simply means figuring out a possible measurement result. In other words, 'prediction' means basically knowing what the result of a specific observation will be.

"Let me give you a simple example. We all know that when we step out on a clear night and see the Moon, we can predict with certainty that we will see the Moon again if we look away for a moment and look up again. There is something there that is an element of reality, and we commonly call it the Moon. A physicist would go even one step further and say that a physical quantity we can predict is the position of the Moon in the sky."

"But," Alice interrupts, "no one can predict this position with certainty. There is always some uncertainty to it. We can never tell exactly where the Moon is."

"Fair enough," John replies. "That is correct, and it points at an important feature of any observation in physics. That is, whenever we perform a measurement of any quantity, there is always some small uncertainty to it. No measurement instrument is infinitely precise. So if we want to predict the position of the Moon, we have to measure its position as well as possible at a particular time and calculate the future positions from that information."

"Actually," Bob says, "I remember that we have to know the position of the Moon, the Sun, Earth, and all the planets to figure out precisely where the Moon will be in the future."

"Yeah," John nods, "this is correct, and there is always some error in this kind of measurement. Therefore the position of the Moon cannot be predicted with infinite precision, but it can be predicted well enough for space travelers to go up there. We can calculate precisely enough the trajectories of their spaceships so that they really arrive. But, to make a long story short, prediction with certainty does not mean prediction without measurement error."

"Now," Alice asks, "why is this just a sufficient criterion, as you said before, and not a necessary one?"

"Well, clearly," John answers, "there can be cases when an element of reality exists, but you cannot predict the value of a physical quantity with certainty. It could be that we are quite uncertain about something happening in the future, or even worse, it could be that we don't know anything about some future event. For example, we all might meet someone in the future who will become important in our lives, but we have no idea now who that person will be. The person is already an element of reality. The EPR criterion could not cover that case. But that person, if she or he is of our age, is already alive, so that person exists already, is already an element of reality. But Einstein, Podolsky, and Rosen were sure that if you can predict something with certainty, you can reasonably assume that there is an element of reality associated with it."

REALITY IN ALICE AND BOB'S EXPERIMENT

"But what does all that have to do with the experiment Alice and I are doing?" Bob asks. "Where can we apply this 'prediction with certainty'?"

"OK, let's try to apply this EPR reasoning about an element of reality to your experiment," John answers. "On the first day, you found out that the individual results of your measurements are completely random. Independent of the settings of your respective switches—plus, zero, or minus—the red and green lamps flickered with equal probability. So one would conclude that an element of reality following the EPR definition cannot be assigned to the individual measurement result, but it might still exist."

"Yes, that's what we found out on the first day. And maybe you just explained to us why we were so frustrated!" Alice exclaims, smiling. "It was the fact that we were not able to tell whether some reality was behind our observations or not. But all that changed on the second day."

"Yes," Bob agrees, "on the second day, we discovered that our results are perfectly correlated. Once we chose the same setting for our switches, both of us got the same result, red or green. But I don't see the connection to the EPR criterion of reality." Bob frowns.

They sit together quietly for a while. John does not say anything, since he wants them to find out by themselves. He just gives them a hint. "Just consider what it means that you get the same result on the other side once you know one result."

"I've got it!" Alice says suddenly. "Once I know my result, I can make a perfect prediction of what Bob will get if he has his switch at the same position as mine. Say my switch is on plus and I get green. I can predict with certainty that Bob's green lamp will also flash if he also has his switch on plus at that very moment."

And Bob adds, "You're right! So we can apply the EPR definition of elements of reality here, because either of us can predict with certainty what the result of the other one will be. Everything's clear now!" Bob slaps his head. "Why didn't we figure that out earlier? If my green lamp flashes, I know that Alice's green lamp will also flash."

"*If* I chose the same position for my switch as you did for yours," Alice adds.

"But wait a moment!" Alice suddenly exclaims. "I'm getting an idea about the stuff we are really playing with in our experiment. We know that the stuff produced in the source is sent to our two laboratories along glass fibers. So it must be pulses of light that are sent along from the source, pairs of pulses. But since Professor Quantinger told us that we are doing a quantum experiment, the pulses must actually be just individual photons," she concludes triumphantly.

"You're right!" Bob says excitedly. "The source must be producing pairs of photons and sending one photon of each pair to you and one to me. From that model, we can easily understand why we register these twins. But we also can understand why we do not always register twins. The photons might simply get lost along their way, or the detectors aren't perfect and don't always register the photon when one is coming in. So some get lost, and some don't get registered."

"Bingo!" John exclaims. "You guys just took an important step toward understanding what you are doing. I knew it. I was always confident that you would discover what was going on here. Why don't we go down to the basement and have a look at the source? I am now allowed to show it to you."

"Super!" Alice and Bob answer. "We need to move around a little anyway."

In the basement, John opens the box, and indeed, here it is (Figure 23), the source of photon pairs. John explains all its parts.

"Looks rather simple," Alice says.

"It is simple," John explains with some pride. "But to make it that simple, many years of experimental development were necessary. Since you know now what the source looks like, all you have to do is to explain the correlations of the measurements on both sides. For example, what do you think could be the reason why you always get the same results when you choose the same switch position on both sides?"

"Well," Alice says, "for each photon we have two possible results, red and green, and we have three switch positions. Which property of a photon could it be that has two possible values?" She scratches her head.

"I've got it!" Bob exclaims. "Do you remember the professor's lecture on the polarization of light? We learned that each photon can be vertically or horizontally polarized. So maybe what we do is just measure the polarization of the photons. One of the two results, say, red, corresponds to horizontal, the other, green, corresponds to vertical."

"Again, you've got it," John responds. "But now you have to find out what the positions of the switches mean."

"Well," Alice muses, "in the lecture, the professor showed us how you can change the orientation of a polarizer and get very different measurement results. So maybe the three positions of the switch—plus,

Figure 23. A source of entangled photons. A laser (*top*) creates a blue beam that enters a crystal in the center of the figure. There, pairs of entangled photons of red color are created. These are redirected and finally coupled into glass fibers (*bottom right and left*).

zero, and minus—correspond to three orientations of a polarizer that each of us has."

"Right again!" John says. "Zero corresponds to zero degrees, plus corresponds to a rotation to the left by 30 degrees, and minus corresponds to a rotation to the right by 30 degrees."

"And it must be polarizing beam splitters with two outputs," Bob says, "because we get two results." He goes to the blackboard and makes a small sketch (Figure 24) of the principle of the arrangement with the source and the two polarizers on his and Alice's sides.

"Apparently, the photon pairs are born with the same polarization and therefore give the same result, red or green, horizontal or vertical, on both sides when the same orientation of the polarizers is used, that is, when we measure the same kind of polarization."

"So we finally understand what is going on," Alice says. "Our project is finished."

"Yes," Bob agrees, "and if your polarizer is not at zero, but rather at plus or minus, there won't be a perfect correlation between our measurement results. But I can even predict the probabilities with which your plus and minus detector will flash. So basically, we are now able to explain our results," Bob finishes triumphantly.

"Sure," John replies to Bob, "with your polarization model, you can now try to explain all your results. All you have to assume is that one measurement, a measurement of one photon, gives some result, namely, its polarization, horizontal or vertical, for the respective orientation of the polarizer chosen. Then you know that the other photon has the same polarization. And if Alice happens to have her polarizer in the same orientation, you can predict with certainty her measurement result. You can predict with certainty which of the two lamps, red or green, will flash. If she happens to have chosen the other orientation, you can at least predict the probability with which each of the two detectors will flash."

"So, we're done!" Alice says. "The experiment is finished, right?"

"Yes!" Bob exclaims. "We've found out that the source is very simple. It emits pairs of photons that are either both horizontally polarized or both vertically polarized. If Alice and I have the same orientation on our polarizers, we get the same result, and if we have different orientations, we don't get the same result."

John smiles widely. "This is a model that was discussed very early. It

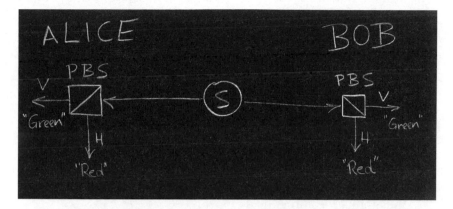

Figure 24. The principle behind Alice and Bob's experiment. The source (S) sends out pairs of photons. Alice and Bob measure their polarization with polarizing beam splitters (PBSs). One of the results, horizontal (H), triggers the red lamp; the vertical (V) result triggers the green lamp. Also, Alice and Bob can change the orientation of each polarizer by rotating it around the incoming beam.

is called the Furry hypothesis, after the American physicist Wendell Furry who actually disproved it, who actually showed that it is wrong. The idea was put forward as a possible suggestion, a possible way out, by Schrödinger."

"That idea is wrong?" Bob asks. "How could it be wrong?"

"Well," John answers, "you have to find that out by yourselves, because I have to rush off to meet Professor Quantinger. He gave me an appointment to talk about my Ph.D. dissertation. And, by the way, the EPR story is not finished yet."

Bob shouts after John, "You can't leave us like this!"

John, over his shoulder, shouts back, "You'll find out! Just think carefully about all the measurements you have done so far. They will allow you to see why your own model is wrong."

And around the corner he disappears.

Alice looks at Bob. Bob looks at Alice. They scratch their heads. Neither of them has any idea how to proceed, and there is no point in going back to the laboratories, since John has told them that they already have all the data they need. They go to get a cup of coffee and sit down together with all the data.

"John told us that with our picture of polarization, we cannot explain everything. If we set one polarizer at zero, we will get either horizontal or vertical, and we know that the other photon is also horizontal or vertical. We know for sure what the measurement result on the other end will be if the polarizer there is also set at zero orientation," Bob remarks.

"So," Bob continues, "there must be some conflict between our polarization model, which John said is wrong, and the idea that we can explain everything using polarization, which John said is correct. This is really weird. What is going on here?"

"I don't know. But let's consider very carefully what's happening," Alice says. "Suppose I set my polarizer at zero, then half of the photons show horizontal polarization, half of them vertical polarization—horizontal and vertical along the zero-degree orientation.

"So this is a stream of photons," Alice continues, "where each one is horizontally or vertically polarized. And an identical stream of photons goes over to your setup. If your polarizer is also set at zero, then exactly those that show up horizontally at my end will show up horizontally at your end, and the same is true for the vertical polarization."

"Therefore," Bob concludes triumphantly, "the model is correct. It explains the perfect correlations in a fantastic way."

"Maybe," Alice goes on, "we should also look at the other cases where we got perfect correlations. Suppose I set my polarizer at plus. Then half of the photons will show horizontal polarization along the direction rotated by 30 degrees from the previous one, and some others will show vertical polarization with respect to that direction. And then, I know that your photons will be polarized horizontally or vertically the same way. So let's draw a sketch here [Figure 25]. Again, we can explain the data perfectly," Alice finishes.

"I can't see the problem yet," Bob says.

"Neither can I," says Alice, "but maybe we could also analyze the minus 30-degree case."

"That's ridiculous," Bob answers. "Minus 30 degrees is the same as plus 30 degrees, just rotated."

"Nevertheless, let's think about that and let's make another graph" (Figure 25).

"I'm sure the solution is staring at us. We just don't see it. Something must be wrong with this picture. What could it be?"

"I got it!" Alice suddenly exclaims, and she jumps up from her seat. "How does the source know what to emit?"

"What do you mean, how does the source know what to emit? The source emits photons of equal polarization."

"Yeah, but oriented along which direction? Are they horizontal or vertical for the zero-degree orientation, for the plus 30-degree orientation, or for the minus 30-degree orientation?"

"That should be the same," Bob says.

"No, it's not the same!" Alice says. "Look at the picture! The three pictures [Figure 25] are very different, and all that is different between the three is the polarization orientation of both photons. How can the source know beforehand what the orientation of the polarizers is and therefore what kind of photons to send out? And even worse, suppose I set my polarizer at zero degrees and you set yours at plus 30 degrees. Since both our stations are basically the same, there should be no difference. What is the polarization of the photon pairs emitted then? Is it horizontal or vertical along zero, or horizontal or vertical along plus?"

"I get your point!" Bob gets excited too. "Somehow, the source has

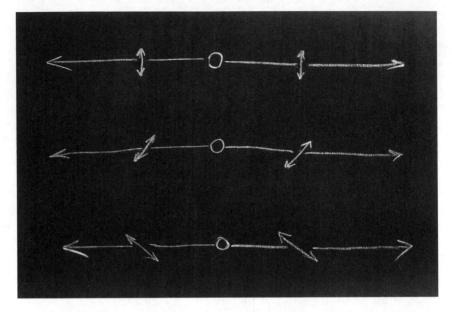

Figure 25. A source in the center emits pairs of photons with given equal polarization, one to the left, one to the right. The two photons can, for example, both be vertically polarized (*top*) or polarized along a rotation of 30 degrees in one direction (*center*) or the other (*bottom*).

to make up its mind what to send out, and it must therefore know what the orientation of the polarizers is and act accordingly."

"But I have a feeling that this cannot be the explanation, because instead of being a couple hundred feet apart, the two of us could be at a much greater distance. And we could actually switch around very rapidly during the experiment. We could switch around so fast that the source would get confused. For example, I could have my polarizer at zero at one moment, so the source sends out photons that are either horizontally or vertically polarized in the zero frame. But then I could switch rapidly to another direction, and suddenly I would receive photons that are not oriented along my axis," Alice says.

"But maybe that does not matter. Maybe the source emits photons that are horizontally and vertically polarized along all possible directions—just a mixture. Let me do a drawing," Bob says (Figure 26).

"Now it's clear why that can't work," Alice continues. "Among all these many pairs, there are only some that have either both horizontal or both vertical polarization. Most of them have something different."

"That's true." Bob continues, "If we have our polarizers oriented at zero and if we both pick out the horizontal and vertical photons that are polarized for zero orientation, then, clearly, both you and I will get the same result. Either we both get horizontal, or we both get vertical. But let's now pick out pairs, for example, where both photons are horizontal-plus polarized, that is, polarized along a horizontal direction rotated by 30 degrees. What will happen if they meet my polarizer with zero orientation? Some of them will go into the horizontal channel, making the red light flash, just as desired. But some will go into the vertical channel, making the green light flash. Likewise on your side. Some of your photons will go into the horizontal, some into the vertical channel. So sometimes your red and sometimes your green light will flash. Will they always flash together?"

"Evidently not, because a photon on one side does not know what the other photon will do. Remember, we have learned from the professor that quantum mechanics gives us only probabilities. So, each photon coming to me will have to make up its mind whether it goes to horizontal or vertical, and likewise, each photon going to your piece of apparatus will have to make up its mind whether it goes to the horizontal or vertical channel. They will do that independently. Sometimes—

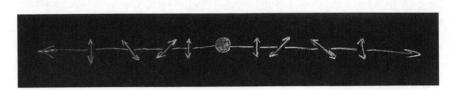

Figure 26. A source emits a series of photon pairs, one after the other. In each pair, the two photons carry the same polarization. But the directions along which they are polarized vary from pair to pair. Such a picture cannot explain Alice's and Bob's measurement results.

actually, in quite a number of the cases—different lamps will flash," Alice replies.

"Bingo!" Bob exclaims. "Our model cannot work, simply because two-thirds of the photons are emitted with polarizations where the measurement is not definite. Since the measurement result is not definite, but probabilistic, random, I would not be able to predict with certainty what your result is."

"That's a fantastic result. We should get to John and tell him."

Alice and Bob are so excited they decide to wait at John's office for his return from his discussion with the professor. Soon, he shows up in a good mood.

"Professor Quantinger basically accepted the outline of my dissertation and the couple of chapters I have written already. So it sounds as if I'll be finished and get my doctorate within a few weeks. But I can see from your faces that you two have something to tell me. Come in."

Alice and Bob again settle comfortably in the chairs in John's office. They explain what they have found, describing the model that the photons are born as pairs, each with a fixed polarization. But the polarizations of the two photons are identical. Then they also explain to him how they assume that the source emits a mixture of many pairs, but each pair is polarized along a different direction. They finally explain to him why, in their thinking, the model is not correct, because it does not provide the perfect correlations seen in the experiment.

"This is very good and clear thinking. I congratulate you, Alice, and you, Bob. You are certainly physicists, able to carry a model through clearly to find its consequences."

Alice and Bob are delighted by that remark.

"But," John continues, "you have a challenge now. We still would like to explain why the two photons give the same measurement result if both of your polarizers are oriented in the same way."

"Yes, that is really amazing," Bob says, scratching his head. "We have seen in the experiment that both photons show the same polarization when they are measured the same way. But we have also seen that the photons do not carry that polarization before they are measured, and if we assume now that the individual measurement result for the individual photon is random, then we have something really mysterious. How is it possible that two random processes separated by large distances always give the same result?"

"That's exactly the point," John emphasizes. "This was worked out very clearly by Schrödinger in his 1935 reply to the EPR paper. For him, it was a really surprising thing that we can make perfect predictions for joint measurements, but the individual measurement cannot be predicted perfectly. There is an element of randomness. This, he thought, is something that is only possible in quantum physics and not anywhere else. Schrödinger coined the term 'entanglement' for this situation, and he called entanglement *the* essential feature of quantum physics."

Alice replies, "Well, couldn't it be that when I do my measurement with my polarizer at some defined orientation, somehow my apparatus tells Bob's apparatus what is being measured?"

THE LOCALITY ASSUMPTION

John smiles. "Einstein, Podolsky, and Rosen were aware of that question. They formulated what is now called the locality assumption. It goes as follows: since at the time of measurement the two systems no longer interact, no real change can happen in the second system as a consequence of something that is done in the first system."

"Again, heavy stuff," Bob reacts.

"But no problem," John says. "Let me explain in more detail. Their argument goes roughly as follows, applied to your experiment. At the moment of measurement the two physical systems, the two photons, which were created at the source and sent to your respective measurement stations, are no longer interacting with each other. They are widely separated. In fact, we can easily imagine that they are so widely separated that it takes any signal or any sort of information a long time to go from your, Alice's, to your, Bob's, measurement stations. We know that the speed of any signal is limited by the speed of light."

"So, there cannot be any information telling Bob's station what I measure. That's quite exciting!" Alice exclaims. "So that explanation is out."

Bob doesn't give up as easily. "How do we know that it really works that way? Our two measurement stations are very close to each other, so a signal could arrive in time to tell my apparatus what you measure."

John interjects, "Yes, in principle that could still be possible for your experiment. But there was an experiment by Gregor Weihs and his colleagues at Innsbruck University a few years ago that definitely ruled out that possibility. But let's talk about that later and now just focus on what we can do with this new knowledge."

"OK," Bob agrees. "I believe you that our experiment will work over any distance and that it cannot be fixed by having a signal go from one station to the other. But then, we have a huge problem. How can we explain the perfect correlations we observed if our apparatuses happen to be oriented in the same way?"

"Why is that a problem?" John challenges him.

"Well, maybe it is not a problem, but it's an open question," Bob answers. "If I remember the EPR argument about an element of reality correctly, then we can apply it to our experiment. If I do a measurement with any of the orientations of the polarizer, I can predict with certainty what the measurement result on Alice's side will be. She just has to orient her polarizer in the same way, and she can prove that my prediction is correct."

"Quite right," Alice continues. "I see what the point is. It is therefore reasonable to assume that there must be an element of physical reality carried by Bob's photon that makes the green lamp on Bob's side light up, and not the red one, if my green lamp flashes and his switch is set the same way as mine. So there must be something on Bob's photon that makes Bob's measurement apparatus behave in a way that the green lamp lights up. This must be some property of the photon. It might be a property that can easily be seen, but we know it cannot be the polarization. Or it might also be something hidden, something that is very difficult to see or maybe even impossible to observe directly."

Bob, understanding Alice's suggestion, jumps up. "Oh, I can see a very beautiful analog. We can compare this with the genes determining the features of a person's body. Whether someone has black hair or blond hair is determined by her genes. So the picture is really like the story of the twins. If we pick out a specific pair of twins, the two have black hair because they both carry the same gene determining the color of their hair."

Alice continues, "Yes, we can look at it that way. For example in the case when both switches are set to zero, we might assume that the particle carries a zero feature that determines whether the green or red lamp lights up when the switch is set that way. We can certainly argue that such a feature must exist, because whenever the two switches are set the same way, the same lamps light up. Or in other words, the two photons show the same polarization, either horizontal or vertical."

"What you just introduced is called hidden variables. So, let me wrap the hidden variable idea up for you," John suggests. "Just consider two boys who are identical twins.

"First, we see that the twin boys, as far as we can tell, are identical. They have identical features.

"Second, by considering their development, we realize that their features were identical from the beginning. Our twin brothers were already born with the same color hair, the same color eyes, and so on.

"Third, we learn a very simple explanation of their identity in that both twin brothers carry the same genes. They carry the same cellular information.

"These genes can be seen as hidden properties of our twin brothers. Indeed, these genes are contained in every cell of the body. They were long unknown to us until modern biological research discovered them. So, these hidden properties determine the features of both individuals. The development of the individual features is then a consequence of the specific information laid down in the genes. For identical twins, the specific information is the same, and it leads to identical properties for the two persons.

"But the individual properties are not completely determined by the genes. They are also influenced by the environment, starting early on with slightly different conditions in the womb. For example, even identical twins have different fingerprints. The question of how much individual features depend on the environment and how much they are influenced by the genes is still a point of strong debate in science. But for our discussion, this is not important.

"It's tempting to try to use the same explanation for your quantum experiment," John concludes. "There, identical features of the two photons are measured when the same measurement is done on both sides. So, in the case of twins, the measurement was, for example, looking at hair color. In the case of photons, it is measuring their polarizations for some orientation of the polarizers. So the analogy would be perfect if our photons also carried something like genes. Physicists actually did consider this kind of explanation. Interestingly, this story does not work for our quantum twins. Quantum twins are something very different from classical twins! The identity of entangled quantum pairs cannot be explained through hidden properties, as we will now see in detail."

JOHN'S STORY ON LOCAL
HIDDEN VARIABLES

"We have now arrived at the discussion of what physicists call the model of local hidden variables. The genes in biology are such a model," John continues.

"The basic question is whether the measurement results can be explained by unknown features the particles carry. So, for example, if both your polarizers are set at zero degrees, the argument goes that each particle carries instructions that tell it whether it should show horizontal or vertical polarization, and thus make the green or the red light flash. Such variables are called hidden, because we can't necessarily see them directly. It's enough that they act in such a way that the correct measurement result shows up. And we call them local because the result on, say, Alice's side is independent of whatever Bob does on his side. The result depends only on the local settings of her apparatus and on the hidden variable her particle carries.

"Now, an important point is that you, Alice and Bob, can choose the orientation of the polarizers at any time you wish. Actually, you can change it at the last instant, after the photons have already left the source. This has the very important consequence that such hidden variables, such elements of reality, must exist for all three settings of the apparatus—plus, zero or minus—because no matter which of the three settings is chosen by Alice, she can predict with certainty that the same lamp she observes on her side will light up on Bob's side, should his switch be set to the same position she chose. So, whenever she chooses the plus position and observes, say, the red lamp lighting up, she can predict with certainty that Bob's red lamp will also light up, should his switch also be set to plus. In other words, both particles must be prepared to give a well-defined result for any of the polarizer orientations.

"This is related to a very important point, which Einstein, Podolsky, and Rosen call attention to. The two measurement stations, Alice's and Bob's, could actually be separated by much larger distances. Say, in the extreme, they might be light-years apart from each other: one measurement station on Earth, the other one on some distant star, and the source in between. Such an experiment has not been performed yet, but there is no reason why it should not be performed someday, and the results, according to all we know today, should be the same."

"So, here we meet the EPR locality assumption again," Alice intercedes.

"Quite right," John answers. "The element of reality predicted by you, Alice, must be completely independent of whether Bob happens to choose the same setting of his switch as you or not. Even more, whether or not the system carries this additional element of reality must be completely independent of whether Bob cares at all to do the experiment.

"In the same way, our two particles must carry the same hidden properties determining which lamp they will light up for each of the three settings—plus, zero, or minus—independently of whether we actually care to look at them. In a simple way, as my colleague Mike Horne once pointed out, we can imagine each of the two particles carrying a list of instructions that tells the particle what to do if it meets a polarizer that happens to be oriented in a particular direction. Clearly, the particle has to carry an instruction for each of the polarizer orientations it might need. So, if we restrict ourselves to plus, zero, and minus, the list could be like this—"

John writes a list on his blackboard (Figure 27, *top*).

"The next particle could have the following list"—and he writes the second line on the blackboard (Figure 27, *center*)—"and the third one this one [Figure 27, *bottom*] and so on and so on. Each particle has such clear instructions."

"Yes, and the instructions are obviously the same for the two particles," Alice remarks, "and each particle goes on its travels with its list."

"Yes, sure," Bob says, picking up the line of argument, "and when a particle meets a polarizer, it checks the orientation of the polarizer. Then, it looks at its list of instructions to find out which detector it has to go to in order to be registered."

"That is," John continues, "a very simplified way of representing things. But in principle, hidden variables work like that. Each particle

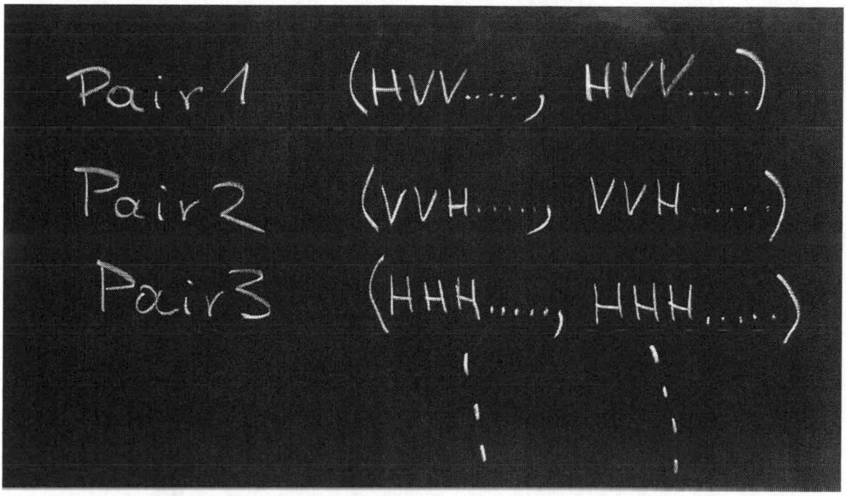

Figure 27. List of instructions for three photon pairs with the goal of obtaining perfect correlations. Each photon carries instructions for every possible orientation of the polarizer. The instructions say whether the photon should be polarized horizontally (H) or vertically (V) for that orientation. Within the parentheses, the instructions before the comma refer to the first photon of the pair, the instructions after the comma to the second photon. The second photon always carries the same instructions as the first one. But the instructions vary from photon pair to photon pair created by the source. The first line, say, means that both photons will show horizontal polarization when meeting a polarizer oriented at plus 30 degrees and they will show vertical polarization when they meet a polarizer oriented at zero degrees or minus 30 degrees.

carries properties that define which result it has to show at which kind of measurement. The perfect correlations then are simply explained by the fact that the hidden variables of both particles are identical. And since it's not clear from the beginning which measurement will be performed, all particles have to carry instructions for all possible measurements."

"A very plausible model," Alice remarks. "A little complicated, but it should work."

"The bad news is," John continues, "and at the same time the really exciting point of the whole story we are investigating is that these considerations do not work for entangled quantum particles even though they work very well for human twins. John Bell found out that when you calculate what this kind of model predicts, it does not agree with quantum mechanics for all possible measurements. The details we must keep for another time. I just want to tell you that, since this model explains the perfect correlations that you got. It explains the cases when both of your polarizers are oriented in the same way. But it does not explain all possible correlations." With these words, John concludes his presentation of the model.

After John's lecture, there is silence for some time in the room. Then Alice says meekly, "What kind of game are you playing with us? We had a beautiful model, the model using polarization, and all that happened was that we had to shoot it down. Now, we have a second model, which is even more beautiful than the one with polarization. You are telling us it's wrong. How long will this go on?"

"Don't worry," John reassures her. "You are very close to seeing the light at the end of the tunnel and to learning a very deep lesson about nature. But for now, you are on your own again."

Alice and Bob leave in utter confusion. How could the model of the hidden variables be wrong? Whenever they set their apparatuses to the same setting, they got the same result. How could it be that the same result occurs on both sides, but without the system carrying a property that determines what the result should be? If that is not the right explanation, what could the right explanation be? The only other way to explain the results would be some secret communication between the two pieces of apparatus. But by the EPR locality assumption, such communication can be excluded for widely separated pieces of equipment. If one piece were on Earth and the other one on a distant star, as John mentioned, then it would take years for this information to arrive, since nothing can travel faster than the speed of light. So Alice and Bob are again completely confused and do not know how to proceed.

"How can we find out," Bob says, "what John meant by saying that the model is wrong? Let's give him a call."

When John answers the phone, Alice asks him directly, "Are we supposed to find out that the model is wrong purely by thinking, just like last time?"

"No," John says, "you can do some more measurements."

"But," Alice sighs, "we don't know what to measure now. We've observed perfect correlations if our two polarizers are oriented parallel. And we've observed that no perfect correlation is possible when the settings of the two apparatuses are not the same. So what can we do now?"

John's answer is a little cryptic. "I am not supposed to tell you what you should do now. You should find out yourselves. One hint: you have not yet looked at all the possibilities. Remember, you can choose the settings of your switches freely and you can count photons."

Bob replies that this does not make much sense to him. But John refuses to give them any more hints.

ALICE AND BOB'S EXPERIMENT GIVES CONFUSING RESULTS

So, grudgingly, the next morning, on Thursday, Alice and Bob meet and worry about what kind of measurement they should do now.

"We have done all the possible combinations," Alice starts, "between settings on your side and settings on my side, because that's what we did when we started to play with the apparatus. So what more can we do?"

"Let's recall what we did so far. On Monday, we each measured our photon on our own, and we discovered that there is no order in the data. We then learned that this is because there is no rule for measuring the individual photon. We have complete randomness," Bob remarks.

"On Tuesday we discovered," Alice continues, "that our photons are emitted by the source in pairs, even if we only detect about a fifth of all the photons in our detector."

"And on Wednesday," Bob says, "we saw that when both of us measured the polarization at the same orientation, we got the same result. So we discovered the perfect correlations when we had happened to choose the same settings on our two pieces of apparatus. But wait a minute! We did not really pay any attention to the results we got when we happened not to choose the same setting. Maybe we should look at the numbers there."

Alice protests. "What can these numbers tell us? We found out that they are random. Whenever my red lamp lights, it could be your green or your red lamp if you don't have the same setting."

"Which we now understand, by the way," Bob continues, "because we know that if you set your apparatus at zero—if you really orient your polarizer at zero degrees and if you measure a photon—its polarization is either horizontal or vertical, depending on the channel whose detector is finally triggered. And then, if my polarizer is not parallel, the photon can go into either channel and that's that. But maybe we should look at the exact numbers we get."

"So, in other words," Alice gives in, shrugging her shoulders, "we might as well do that, since there is nothing else we can do."

So they decide to take all possible combinations between the settings on Bob's side and on Alice's side, without choosing the same settings, because they had done that part of the experiment already. So the procedure now simply is to set Alice's switch, for example, to plus and Bob's to zero and to count for 200 seconds how often the combinations green-green, green-red, red-green, and red-red turn up. Just to make sure, they decide to do all these measurements twice.

Each of them ends up with similar lists, as before. There are twelve lists altogether. For each photon observed, there are entries for the time when the computer registered an event, the orientation of the apparatus, and the color of the lamp that flashed. With the printouts of the twelve lists, they again get together at the cafeteria to figure out what was going on.

"So what are we going to do with all this mess?" Alice says.

"Well, maybe we first should find out at which times both of our detectors registered some stuff," Bob answers.

Alice smiles. "We already know that the stuff is photons. So let's talk about photons. That makes it easier for me."

Bob smiles back. "Well, if you know what a photon is, you should tell everyone. Einstein never found out what a photon is. He is supposed to have said toward the end of his life, 'Fifty years of intensive brooding did not get me closer to the answer to the question: What are quanta of light? Today, every rascal thinks he knows it, but he is in error.' Never mind Einstein. Alice knows it," Bob teases her. "But I guess to use the word 'photon' is simply a way to talk about the situation, even if we don't fully understand what it is."

"Fair enough," Alice concedes. "This is probably what it's all about, that we have to have a way to talk about what we see."

The only thing they can do now is to tally the results up according to which lamp lights up on which side together with which lamp on the other side—four combinations, green-green, green-red, red-green, and red-red.

"So, let's look at the plus-zero results," Bob suggests, "those results where your polarizer is oriented at plus 30 degrees and mine at zero degrees. Taking both two-hundred-second measurement periods for that orientation together, we see that we had eighty-nine coincidences. That means eighty-nine times there was a particle measured in both of our pieces of apparatus. In thirty-one cases, we had red-red, in thirty-five cases green-green, in eleven cases red-green, and in twelve cases green-red. What could that mean? All these numbers are different."

They both look at the numbers for a while. Alice breaks the silence: "Well, I have a feeling. If we sum up those cases where both of us register the same color, we have 31 + 35 = 66. And in those cases where we registered different colors, we have 11 + 12 = 23. So, roughly speaking, we're about three times more likely to get the same color on both sides as to get different colors. We should be able to explain these numbers. We know that the two polarizers are making an angle of 30 degrees with each other."

"I get what you mean," Bob continues. "My apparatus was at zero degrees, which means that when I measured the result green, the photon was vertically polarized. Then, given my result, your photon is also vertically polarized."

"Yes," Alice continues eagerly, "and I measure this vertically polarized photon now at an angle of 30 degrees. Remember Malus's law, which we learned in the professor's lecture on polarization? It tells us that such a photon would be found to have vertical polarization with respect to the 30-degree orientation and therefore be observed in your green detector with a probability of 75 percent. And it will be found to have horizontal polarization for my polarizer and be registered in the red detector with a probability of 25 percent. This is just what we observed within the usual measurement uncertainty."

"So we should look at other results," Bob suggests. "Maybe the next one we should look at is the zero-plus experiment, where you have your polarizer at zero and I have mine at plus."

It turns out that the zero-plus experiment is also consistent with Alice's hypothesis. True, the numbers are not exactly 75 percent and 25 percent, but close. They accept these small deviations, because they had learned earlier that when these photons are counted, the numbers are never exact. They tend to fluctuate a bit.

"That means," Bob proposes, "that for all other experiments, we should get the same kind of data: 75 percent equal results—red-red or green-green—and 25 percent different ones."

They check their lists and indeed find this prediction verified for the combinations zero-plus, minus-zero, and zero-minus. But for the combinations plus-minus and minus-plus, the results look quite different. Here, they find 25 percent identical results and 75 percent different ones.

"But that is now easily explained," Alice says, "because there is an angle of 60 degrees between the plus and minus polarizer orientations. And then, Malus's law predicts just what we observed."

Alice and Bob are excited. They are now able to understand the numbers they measured.

"Great data" they suddenly hear from behind. It is Professor Quantinger, who had happened to come into the cafeteria and, seeing Alice and Bob sticking their heads together over their notes, had sneaked up on them. He looks at the piece of paper in front of them and immediately understands what the data tell them.

"Congratulations! You guys did an excellent job. This is exactly what you should have gotten."

"But all this does not make sense!" Alice exclaims.

Bob adds, "In the case where we had the same setting on both sides, we obtained perfect correlations. We then simply assumed that both photons were born with the same polarization. That model did not work, because the source cannot know which polarization to emit. Therefore, the photons do not have any polarization before the measurement. But in the moment of measurement, they have the same polarization. I am really confused."

"And in order to be able to explain these perfect correlations," Alice continues, "we assumed that the particles carried some unknown properties, some hidden variables that tell them which of the detectors they have to trigger. But John told us that this model violates the so-called Bell's theorem, which we don't know yet. And he told us to do further measurements. So we measured the other correlations. But we can't make much sense out of it. Why should that contradict the assumption of these hidden variables carried by the photons?"

The professor sits down.

"You guys really measured all the data you need to to refute the model of local hidden variables. But let me tell you a little more of the story. After the famous paper by Einstein, Podolsky, and Rosen, which you know about already, it took thirty years until the great Irish physicist John Bell found out how significant results of the kind you measured are."

"Yes, John mentioned Bell's theorem. So this is what it's all about?" Bob asks.

The professor answers, "Yes. I'll tell you the whole story now."

JOHN BELL'S STORY

"John Bell [Figure 28] was an Irish physicist working at CERN, the European Organization for Nuclear Research in Geneva, Switzerland. CERN was founded after World War II. That war was a great catastrophe. So after the war, physicists from all over Europe decided to start a new laboratory, where they would get together, exchange ideas, follow their scientific goals together, and collaborate. That way, they hoped to foster mutual understanding. This laboratory was built in Switzerland, which had remained neutral during the war. It contains all kinds of advanced machines, like accelerators, and became one of the world's leading laboratories in physics.

"Bell had studied in Belfast, in his home country of Ireland, and came to CERN in 1960. He was interested in designing new accelerators and helped to make them better and more efficient than before. But on the side, he had always kept a keen interest in fundamental issues. Working on such fundamental issues was not fashionable at the time when the CERN laboratory was built. Such discussions were often considered to be 'merely philosophical,' and many physicists thought that they should be abandoned. They were not *proper*. Another general understanding was that while quantum mechanics may be difficult to understand for the individual physicist, all the big questions had been taken care of already by the giants who had created quantum mechanics, people like Schrödinger, Bohr, Heisenberg, or the Englishman Paul Dirac.

"Bell's worries about quantum mechanics were not really taken seriously by the physics community. Their attitude was, if one really needed to know what the arguments were, one always could go back

Figure 28. Erwin Schrödinger (*top left*) on the Irish coast near Dublin, circa 1942. Albert Einstein (*top right*) in Princeton, 1953. John Bell (*bottom*) during a ride on the Liliput-Bahn (a kind of railway for children in the Vienna Prater area), 1982.

and look up the papers of the founding giants. And this is exactly what Bell did and where he found surprises.

"The American-born British physicist David Bohm had in 1952 written down a hidden-variable theory, a theory going beyond quantum physics."

"We have heard about hidden variables," Alice interrupts. "John told us that they would not work for the correlations, and we haven't yet found out why."

"But what is a hidden-variable theory?" Bob asks.

"Yes," the professor says, "we'll get to the explanation of why it does not work for your experiment. But first, what is a hidden-variable theory? Quantum physics makes only statistical predictions."

"That's not correct," Alice interrupts. "In our case, we get perfect correlations. When I get a certain measurement result, I can always predict with certainty which result Bob will obtain if his polarizer happens to be oriented in the same way as mine. So that's no longer a statistical prediction."

"Excellent!" The professor agrees. "There are exceptions. But for most situations, we cannot really predict with certainty in which detector a particle will end up. So these statistical predictions have always worried people, starting with Einstein. He did not like the random character that came into fundamental theory that way. A deterministic hidden-variable theory is simply a way to overcome this random character. The idea is that each particle carries additional properties that really determine which path it takes, what it does when it meets some optical element, or whether it makes a detector go 'click' or not.

"These additional properties are called hidden variables, because we assume that we cannot directly observe them. We can see only the indirect consequences, because they determine the statistics for many particles. In other words, if I take an ensemble of many particles coming out of some source, the hidden variables might be distributed among them in various ways, but what each particle does is well-defined by its hidden variables. In the end, if we look at many particles, the statistical predictions of quantum mechanics are recovered."

"That sounds like a great idea," Bob replies. "So we would not have to worry about randomness and chance anymore."

"Well, it would be a great idea. But let me go back to my story," the professor says. "Bell, when he heard about the hidden-variable theory that Bohm had written down, he became worried about the situation, since in the thirties, the famous Hungarian-born mathematician John von Neumann had actually proved that such a theory is mathematically impossible. So one of them must have been wrong, either von Neumann or Bohm.

"The problem immediately excited Bell, who sat down to check carefully both von Neumann's paper and Bohm's hidden variable theory. He made two discoveries, both of which were seminal in physics. First, he was able to demonstrate that the original proof by von Neumann was simply wrong. Von Neumann, a great mathematician, had made assumptions that are unfounded in physics. We don't need to go into what his assumptions were. The American physicist David Mermin once called the assumptions made by von Neumann actually 'silly.'

"Another related story is about an interaction between the Austrian-born physicist and Nobel laureate Wolfgang Pauli and von Neumann. One day, von Neumann very excitedly told Pauli that he was able to prove some important point. Pauli, who was known for his sarcastic remarks, supposedly said, 'If physics were no more than being able to prove something, then you would be a great physicist.' Pauli, even if he was not very polite, had a point. Physics is a lot about intuition, and not just about mathematical proof.

"Anyway, Bell was able to dispel von Neumann's proof and therefore opened up the door to new fields of investigation for possible hidden-variable theories, which might go beyond quantum mechanics. So there was nothing wrong with Bohm's theory. What Bohm had actually said was that the individual particles follow well-defined trajectories. They have both position and momentum, that is, a well-defined speed at each time, just like a marble rolling along. The problem is that quantum mechanics says that each particle cannot have both position and momentum at the same time. This is Heisenberg's uncertainty principle. So the question is, How does Bohm get around that? He assumes that his particles are guided by some additional force, a 'quantum potential,' that makes them end up just at those places which quantum mechanics predicts.

"This all seemed to work nicely, and it explained even the double-slit experiment."

"So, we have it!" Alice says. "This is the explanation for our experiment!"

"That's exactly the problem," the professor answers. "In Bohm's theory, the quantum potential of two entangled particles emitted jointly from a source has a very strange property. It is that one particle depends on the other directly, no matter how far apart they are. In other words, if you, Alice, measure your photon and observe it to be horizontally polarized, instantly, that is, faster than the speed of light, that act of measurement changes the quantum potential over at Bob's end. This nonlocality of the potential is something that most physicists even today do not like, and therefore they don't accept Bohm's theory. There are other reasons not to accept it, which are of a more technical nature, but that, again, should not worry you.

"So Bell, having found out that von Neumann's proof does not work and having seen Bohm's nonlocal hidden-variable theory, asked himself whether a local hidden-variable theory would be possible in principle. That would be a theory that worked without this instant non-locality.

"Bell started from the EPR paper. He wanted to know whether a theory that followed the assumptions of EPR was mathematically possible. His paper had the title 'On the Einstein-Podolsky-Rosen Paradox.' It appeared in 1964 in a newly founded journal with the name *Physics*, which was published in New York. Actually, the journal had a very short lifespan and passed away after only one year.

"In his paper, Bell starts from the perfect correlations and the EPR reality criterion that make him introduce hidden variables. Applying then the EPR locality assumption, he proves that any theory built on these ideas contradicts the predictions of quantum physics. Just consider your experiment. Your results are in agreement with quantum physics, but according to Bell, the numbers you found cannot be explained. Your model, where each particle carries instructions about whether it should be horizontally or vertically polarized for any possible measurement direction, is just the kind of local realistic theory excluded by Bell."

Professor Quantinger's eyes focus on Alice's and Bob's lists, and he adds, "By the way, there is one more point about your data. You just have numbers now—numbers of events. Are you able to see the pattern? Clearly, the data have some scatter to them, because the source might sometimes emit a few particles more or a few less by accident and the number of different results on both sides might be a little bit too high or a little bit too low. The same goes for the number of identical results. But can you see what numbers you would get on average?"

Alice smiles proudly. She stops twirling her lock of hair, which has by then become a tight little braid, and her face lights up. "Yes, we've seen that! There is a pattern."

Bob eagerly interrupts. "About three-quarters of the results are of one sort, and one-quarter are of the other sort."

And they show him their calculation of the percentages based on Malus's law.

"Marvelous work!" the professor says. "The ideal results are just what you found. I know this because I know the physics of the source. In the case of both plus-zero and zero-minus, 25 percent of the coincidences should be different and 75 percent should be the same. And in the case of plus-minus, it's just the other way around—75 percent the same and 25 percent different. And in the case of identical settings, the results should be zero different and 100 percent the same. Over there, I see your results for the cases of identical settings."

Bob continues, "So, the 25 percent and 75 percent are, according to Bell, in conflict with the philosophy of local realism?"

"Yes!" the professor exclaims. "And it is absolutely fantastic that a philosophical position can be ruled out by experimental observation. The American philosopher-physicist Abner Shimony once remarked that he is happy to see it to be possible to exclude philosophical positions through experiment. That is the case now. People have done experiments of your kind, and their results clearly exclude local realism.

"But certainly, it is very difficult for you to see by yourself how the contradiction arises. Bell himself once wrote a paper for a broader public with the title 'Bertlmann's Socks and the Nature of Reality.' In that article, he starts from the observation that his friend and fellow physicist Reinhold Bertlmann from Vienna always wears socks of different color [Figure 29]. If you see one of his two socks and it is, say, pink, then you know that the other sock is not pink. The explanation is simply that these are the socks Bertlmann put on in the morning. If they were quantum socks, they would not have any color before they were observed but still always be different.

"So I also wrote a small paper for philosophers. I have it in my office. There I give a quick derivation of a simple version of Bell's theorem. The argument essentially follows a suggestion by the Hungarian-

Figure 29. The Viennese physicist Reinhold Bertlmann always wears socks of different colors. If you see one, then you know that the other one will certainly be different. If these were quantum socks, they would get their color only at the moment of observation, but we know that these are ordinary socks. The colors are different because that is how Dr. Bertlmann put them on in the morning. This drawing was made by John Bell himself, with whom Bertlmann collaborated for many years.

born American physicist Eugene Wigner, also a Nobel laureate. Wigner was one of the few who did not dismiss Bell's theorem as irrelevant when it came out, but who really became excited about it. He found a mathematical proof that is much simpler than the one originally given by Bell himself. Maybe you can stop by my office sometime, and I'll give you a copy. If I am not in, my secretary can give it to you."

But Alice exclaims, "We are just undergraduates! We can't understand complicated mathematical derivations."

The professor calms them down. "The proof is completely without mathematics. I wrote it once for a meeting of philosophers. Since I'm not a philosopher myself, the philosophical arguments certainly are not very complicated either. And actually there is a small improvement in the present version over an earlier one, which is due to a reader of the German version of the paper. This reader is a businessman who certainly has no physics background. But he showed me a way to simplify the central point of the argument. So I am sure you, Alice and Bob, can follow the paper. And more important, the paper will give you a simple result that you can compare with your actual data."

In the afternoon, Alice and Bob go to Professor Quantinger's office and each get a copy of his little paper (see Appendix).

ALICE AND BOB FIND OUT THAT THINGS AREN'T AS THEY THINK THEY ARE

After reading Professor Quantinger's paper (see Appendix), Bob puts it down with a sigh.

"This is quite interesting, but what does it have to do with our work? Where is the connection? We are not looking at twins!"

Alice is more optimistic. Thumbing through Professor Quantinger's paper, she turns to where it shows Bell's inequalities for photon pairs. "Here we have all we need," she explains. "It says that the number of pairs where the first photon exhibits polarization H and the second photon exhibits polarization H' is smaller than or equal to the number of pairs where the first photon exhibits H and the second photon exhibits H'' plus the number of pairs where the first exhibits H' and the second exhibits V''.

"Yes," Bob continues. "This is Bell's inequality, and in the paper Professor Quantinger derives it by just assuming that the individual measurement results are given by local hidden variables. All we have to do now is to identify H, H', H'', and V'' with the appropriate settings and results in our experiment."

"Maybe," Alice says, "we should simply identify H, H', and H'' with the red-light result and V'' with the green light. And have, say, the polarizer orientations for H, H', and H'' (V'') correspond to the switch positions at plus, zero, and minus, respectively."

Alice continues: "Then it's sufficient to consider three different coincidences. First, those between red on my side and red on your side with my switch on plus and yours on zero. Second, those between red on my side and red on your side with my switch again on plus but yours on minus. And third, those between red on my side and green on your side, now with my switch on zero and yours on minus. According to Bell, the first kind of coincidences must always be smaller than the sum of the other two."

Bob eagerly puts numbers in. "We learned that the first kind of co-incidence occurs in 75 percent of the cases, and the second and third each in 25 percent. According to Bell we should have"—and he writes down:

$$75 \leq 25 + 25$$

"This is wrong!" they both exclaim gleefully.

Bob says, "We have finally shown that the locality assumption is wrong. My results depend also on what you do to your photon and vice versa."

Alice contradicts, "No, we have proven that reality does not exist without our first performing an experiment."

"Actually, there are other possibilities."

Alice and Bob turn around on hearing once more the voice of Professor Quantinger, who was strolling up to see what the two were up to.

"One other possibility would be that counterfactual definiteness does not hold. That is, it does not make any sense at all, not even in principle, to talk about measurements not performed. So, if you measure one photon with your polarizer oriented in one direction, it does not make sense, according to that point of view, to talk at all about what the polarization would have been if you had oriented the polarizer in a different way."

"But that would be extremely crazy!" Alice says. "Why shouldn't I even talk about that?"

"Welcome to the club!" Professor Quantinger replies, smiling. "These are exactly the questions that physicists are pondering. Physicists and philosophers have been brooding over these questions ever since Bell's discovery, and the final verdict is not in yet."

"So," Alice says, a little confused, "we have found that Bell's inequality is violated. So what now? The conclusion does not seem to be obvious."

"Well, you have found out a lot," Professor Quantinger replies. "Maybe you want to try to formulate it."

"We have found out," Alice twirls a lock of her hair, "that it is wrong to assume that each photon has its specific property, its polarization, before it is measured."

"Quite right," the professor says, "but there's even more."

"Yes," Bob interjects, "we have established that another explanation is not possible either, namely, that the photon somehow knows what to do if it is measured, that the measurement result is somehow determined by a property of the photon itself."

"Very good," says the professor, "but that in itself would not really be worrisome. There is one important reason why this should bother us."

"A philosophical reason?" Alice asks.

"No, it is a simple experimental result that you guys obtained in the beginning."

"Oh yes, the perfect correlations! We found out that whenever we measure both photons in the same way, they always show the same result. The two are identical."

"Yes," the professor concludes, "that is exactly the point. How in the world can it be that two measurement setups should show exactly the same result when they measure the same feature on a system, if at the same time we know that the assumption that they carry any instructions, or any information about why they should give this specific result, is wrong? This is the real mind-boggling puzzle."

They quietly sit for a while, with Professor Quantinger enjoying the feeling of the minds of the two young people churning, trying to digest what they have just learned.

"For me, the only possibility is," Alice says, "that things are somehow connected in a spooky way."

"But the other alternative is also quite mind-boggling," Bob adds, "that reality does not exist unless we actually observe it. If that is true, it would mean that somehow, the world depends on us, that it somehow depends on whether or not we observe it."

"Or on someone else," Professor Quantinger suggests. "The Anglican bishop George Berkeley formulated this in 1710 with a very brief statement in Latin, *Esse est percipi*, 'To be is to be perceived.' Berkeley actually considered this a proof of the existence of God, God as the ultimate observer, the one who observes the world even when there are no people around."

"But I guess I am not forced to accept that conclusion," Alice pipes up.

"Certainly not," Professor Quantinger responds. "Whatever conclusion you draw is up to you. And if you discover the philosophical meaning of all this, and if you are able to convince others, you can become very famous. But I guess this won't happen in the next half hour, so let's call it quits for today."

FASTER THAN LIGHT AND BACK INTO THE PAST?

We have learned that nothing can travel faster than light. This was discovered by Einstein in 1905 when he proposed his famous special theory of relativity. Einstein's basic observation was that space and time cannot be separated from each other. They form a kind of unity, which today we call space-time. One manifestation of this space-time is that space and time can be transformed into each other depending on how fast one moves. An important point is that inside a spaceship that accelerates faster and faster, time will go slower and slower as you approach the speed of light, which is about 300,000,000 meters, or 186,000 miles, per second (exactly 299,792,458 meters per second).

We now ask, Wouldn't it be possible to go faster than the speed of light by just accelerating? One of the consequences of Einstein's theory of relativity is that this is not possible. Actually, to accelerate when you get closer to the speed of light, you need more and more energy, and it turns out that to reach the speed of light, you would need an infinite amount of energy. Therefore, the speed of light limit is considered to be unreachable for any massive body, for anything that has some mass at rest, such as a space traveler, a spaceship, or even a massive particle like an electron or an atom.

So, why can particles of light travel at the speed of light? The point is simply that photons, particles of light, have no rest mass. It turns out that only objects that have no rest mass can move at the speed of light.

Now, even with the unattainability of the speed of light, we could still assume that somehow, it might be possible to jump directly to a motion beyond the speed of light limit. Einstein showed that if it were possible to move faster than the speed of light, one would be able to arrive at a time before one's own departure. This concerned Einstein quite a bit, because it could lead to very paradoxical situations. The most famous one is the killing-your-grandfather paradox. Essentially, one would be able to kill one's own grandfather and thereby create an irresolvable loop of contradictions (Figure 30).

The loop goes as follows: if you were able to hop on a spaceship, travel back in time, arrive at a time when your grandfather is alive, and kill him, then you would not be born. Therefore you would not be able to hop on a spaceship, you would not be able to travel back in time, and you would not be able to kill your own grandfather. That means, you would still be alive and able to hop on a spaceship, travel back in time, arrive at a time when your grandfather is alive, and kill him, which means that you would not be able to travel back, and so on and so on. We arrive at an infinite loop of contradiction.

Therefore, the requirement that nothing can go faster than the speed of light is actually necessary for the consistency of the universe. The universe cannot be in a state of contradiction. Two things cannot really contradict each other. A person clearly is either dead or alive.

If we think about it a little more carefully, what this consistency argument really tells us is that any going back in time is illegal as long as one is able to influence the past in a way that contradictions of the kind just discussed might arise. So, we might speculate a little. Suppose it would be possible to travel back in such a way that no such contradictions can be created? Then, apparently, there is nothing wrong with going faster than the speed of light. So, for example, if the time traveler has no significant interaction whatsoever with the world at an earlier time, there does not seem to be any problem.

More generally, we can extend our argument beyond traveling to the question of signaling. Is it possible to signal faster than the speed of light? Actually, one can show by similar argument, as in our space-traveler story, that if one were able to signal faster than the speed of light, one could actually signal into one's own past. So one could send all kinds of information into one's past and thus arrive at the same kind of logical contradictions that we saw.

For example, suppose there is a huge lottery win of $100 million by a carpenter, and we send this message back in time. Newspapers appear saying that a carpenter will win $100 million the next day playing the numbers 7, 18, 23, 24, 31, 37. Clearly, if these numbers are printed all

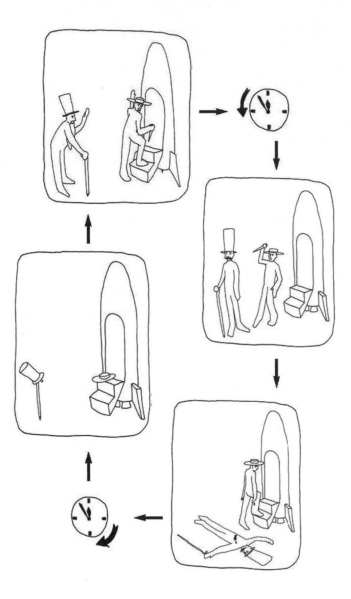

Figure 30. Bob (*top*) says farewell to his grandfather and enters a spaceship. The spaceship then flies faster than the speed of light. Bob arrives back at a time when his grandfather is a young man, long before his father is born, and kills his grandfather. But then, when Bob travels back to the time of his departure, Bob does not now exist because his grandfather died too early. Since Bob does not exist, he is not able to travel into the past to kill his grandfather, which means that his grandfather is alive, and so on and so on. This is a continuous loop of internal contradiction.

over the place, many people will play the same numbers. Therefore, millions of people each will win a few dollars. Among them would be our carpenter. The poor guy would no longer win $100 million (Figure 31). Here we obviously have the same kind of logical inconsistency as discussed before. It must thus be impossible to signal faster than the speed of light.

But we have to be a little more careful. What our rule tells us is that it should not be possible to signal faster than the speed of light in a way that the message can be understood and can change the past. If the message we send cannot be understood, not even in principle, then there is nothing against signaling back in time. Let's suppose the message sent back in time about the lottery win is encoded in some way so that nobody can actually read the numbers. Then there would be no possibility at all of changing the past, because the receivers of the message would not understand it and they would not be able to win the lottery prize. But the message must be encoded in such a way that there is no possibility, not even in principle, of decoding it.

Actually, physicists have discovered some situations where something is able to move faster than the speed of light, but in all these situations, no information is conveyed that might be used to change the past. One case in point is quantum mechanics. Specifically, in some applications of quantum mechanics including teleportation, this is indeed the case.

Figure 31. If we were able to send information faster than the speed of light, then we could also send it back into the past. Therefore, people could read on April 1 the newspaper dated April 2. Suppose the April 2 paper says that someone won a huge sum in the lottery, and it also gives the winning numbers. The contradiction is that then everybody could bet on the same numbers and, therefore, the statement in the April 2 newspaper would be wrong. This again results in a loop of contradiction.

ALICE, BOB, AND THE SPEED OF LIGHT LIMIT

Once more, we join Alice and Bob. They have now finished their experiment and its analysis.

Alice and Bob write a report on the experimental result for Professor Quantinger, who likes the work very much and gives them the best possible grade. He invites them to stop by his office to discuss some of the further implications at any time they would like.

One afternoon, Alice and Bob plan to go sailing, but it starts to drizzle. They decide to call the professor in case he has some time available.

The professor hesitates for a moment, but then makes up his mind: "Well, I am working on a large manuscript, and right now I am actually a bit exhausted and tired. I would love to have a break. Why don't we meet over a nice cup of coffee?"

A few minutes later, rain drizzling outside and coffee steaming in front of them, Alice and Bob are very pleased to again be complimented by Professor Quantinger on their work.

"A few issues remain to be discussed," Professor Quantinger says. "One of them is the question of how fast things happen in your experiment."

After a moment of thoughtful silence, Alice pipes up, "Well, it was my impression that whenever Bob and I have our switches at the same positions, it's always the same light that flashes. So that means that the photons have the same polarization. And these two measurements can happen at exactly the same time if the distances from the source to my lab and to Bob's lab are the same."

The professor says, "That's correct, they are essentially the same, if you allow for maybe an inch or two."

"But then," Bob says, "the situation is quite exciting, because we learned before that none of the photons has its polarization before we perform the measurement. Suppose we set both our switches at, say, plus. At the last moment, the photon decides randomly whether the green or the red light will flash; that is, it randomly assumes horizontal or vertical polarization. And then the other one, no matter how far away it is, will instantly assume the same polarization. And so, the same light, red or green, will flash in the other lab."

"That is exactly the point!" says the professor. "Before either of the two measurements is performed, we cannot assume anything about the photon's polarization. Quite to the contrary, as John Bell has shown us, the assumption that the photon has any polarization before it is observed is wrong."

Alice continues, "Well, even more precisely, it is wrong to assume that the photon, or the apparatus at my place, or both together, might have some property that defines whether red or green will flash independently of which choice Bob makes on the other side."

"Very well remembered, Alice," the professor says.

"But then," Alice comments, "we have a serious problem. This measurement on one side needs to be communicated to the other side faster than the speed of light. Actually, since the distances are the same from the source to Bob's lab and to my lab, the two photons are measured at the same time. So that means that any communication must be instantaneous. This beats Einstein's speed of light limit. So Einstein is wrong!"

"Well, he might be wrong," Bob says, "but we have not really proven that point in the experiment yet. Because what we really do is, we set our switches at some position and count photons for just a while. Then, we set the switch again, and so on. So our changing the switches is very, very slow."

"But," Alice remarks, "why should that be a problem?"

"Well," Bob counters, "in principle, it should be possible for the different pieces of apparatus to communicate with each other. There might, for example, exist a sort of field that spreads from Alice's lab to mine and from my lab to Alice's, and maybe even to the source, that in some way tells the other side what measurement I have chosen to perform. If that information spreads with the speed of light, there is ample time for it to arrive at the other side and for the apparatus, together with the photon, to give the right result. Actually, since our labs are separated by about three hundred meters, or about a thousand feet, it takes only one microsecond for a signal that starts in my lab to arrive at her lab, and we would never be able to switch that fast."

"Well observed!" Professor Quantinger says. "This is a fundamental problem that's already been identified by Bell himself. In his words, 'The settings of the instruments are made sufficiently in advance to allow them to reach some mutual rapport by exchange of signals with velocity less than or equal to that of light.' Bell insisted on a 'timing experiment,' in which the settings are changed during the flight of the particles."

"OK! Let's go back and do that experiment also," Bob declares, having really gotten into the mood of being an experimentalist.

"Well, that's a very hard one," the professor answers. "You need very, very precise and fast clocks to time what happens on each side. But even more important, you need to be able to set your switch extremely rapidly. You need to be faster than one microsecond. This is technically very difficult. And yet, such experiments have been performed. The first experiment was done by a group led by Alain Aspect in Orsay in 1982. In that state-of-the-art experiment, they very quickly switched the photons between two polarizers. The definitive experiment was performed at Innsbruck in 1997 and published in 1999 by Anton Zeilinger with his student Gregor Weihs and their colleagues."

"But how could they change the switch so fast?" Alice asks.

"The trick is simple in principle, though difficult in practice," the professor answers. "As you know by now, by changing the switch in your experiment you just change the setting of a polarizer. Weihs fixed his polarizer in place and positioned a special crystal called an electro-optical modulator in front of it. This electro-optical modulator rotates the polarization of a photon by a certain degree, the degree being proportional to a voltage applied to the crystal. The photon passes through both the crystal and the polarizer. The crystal and the polarizer together are just the same as a polarizer rotated by an angle that can be adjusted."

"So I guess the voltage can be changed very fast, and therefore the polarizer can effectively be rotated very quickly," Bob guesses.

"That's correct," Professor Quantinger says. "So what they did in their experiment was to rotate the polarization within about one nanosecond, which is one-billionth of a second, to a new setting. This was done on both sides independently and randomly, on the sender side and the receiver side—what would have been your and Alice's stations."

"But," Bob says, "the decision to rotate still has to be made by Alice or me at a much earlier time, and there's still ample time for this decision to be communicated to the other side."

"Quite correct," Professor Quantinger answers. "The decision of which polarization to set must also be made very quickly in an unpredictable manner. Weihs used a quantum random-number generator. He took a beam splitter and pointed a weak light source at it. This beam splitter produced random numbers very, very fast, much faster than the one microsecond necessary to communicate from one side to the other. So on both sides, the polarization settings were changed very rapidly. If a photon happened to come, it was registered with a certain setting, and then the setting was immediately changed again and again.

"The important point is that both sides operated completely independently. Sometimes, the two photons happen to be measured with the same settings on both sides, and at other times with different settings. Of course, Weihs had to keep a very precise record of what setting was chosen at which time and whether a photon was registered or not. Long lists of data were produced and compared. All these data contain the kind of information that you found by switching your polarization slowly. Weihs and his colleagues observed that both sides produce the same result when the settings are equal. For different settings, the result was a violation of Bell's inequality."

"That way, any signal that may spread at the speed of light is excluded as a possible explanation," Alice concludes.

"Quite right," the professor continues. "Weihs was able to close what is called the communication loophole. They were able to prove that if there was any communication between the two measurement apparatuses, it had to be more than ten times faster than the speed of light. The exact number is not very important here. It is enough to say that it would be faster than Einstein's speed of light limit. More recently, an experiment was performed by a group in Geneva, led by Nicolas Gisin, that showed this speed to be many times the speed of light."

"This is very impressive," Bob says. "But what does the experiment really confirm? In my opinion, it confirms that if there was communication, it would have been faster than the speed of light. But Einstein tells us that communication faster than the speed of light is not allowed. What is going on here?"

The professor answers, "From a purely logical point of view about the role of information, you can now take different positions. One position is that there is no communication going on. This would have deep implications about the nature of reality and the role of information. We will come back to that later. The other position is that there is communication where the information travels faster than the speed of light. This would have deep implications about the nature of space and time. In any case, we have to find out whether this communication faster than the speed of light is anything to worry about, whether it falls under Einstein's verdict or not."

"But it must!" Alice asserts. "It must run against Einstein, who told us that nothing can move faster than the speed of light."

"Well," Professor Quantinger says, "Einstein told us that things cannot move faster than the speed of light if that would lead to some kind of contradiction, such as that of killing your own grandfather. You would also arrive at this kind of contradiction if you were able to send a message back into the past because, for example, the message could be the lottery numbers that were drawn yesterday evening. If we sent them back into the past, like to the day before yesterday, you could win the lottery and change what happens today, which would change what you would signal back into the past, and so on."

"I see," Bob says. "So we have to check whether we can use entanglement to send information back into the past."

Alice adds, "So all we have to do is send a real signal faster than the speed of light."

"Excellent," the professor replies, "you are quite correct. As before, if we travel faster than the speed of light, we can travel into the past. We can also signal into the past if we have a signal that travels faster than the speed of light."

"But isn't that exactly what we can do with our entanglement?" Alice asks. "In our experiment, whenever Bob and I chose the same setting on our switch, the same lamps flashed, and on each side, it was completely undecided beforehand which lamp would flash. My photon did not know which channel it would go into. Then it randomly went into, say, the upper channel so that the green lamp lit up. And then on Bob's side, the photon did the same. So, Bob's photon must have learned from my photon what to do. Otherwise, how could the whole thing work?"

"Oh, I've got it!" Bob continues eagerly. "I turn my switch, and in the last instant, I set it on, say, the zero position. Then, if my green light flashes and if Alice's switch is also set on zero, I know she also has green. So that's the message. I sent her green instantly, beating Einstein's speed of light limit."

"But where is the signal?" Professor Quantinger asks. "Suppose you met in the morning and thought about having lunch together later. And you agreed that at exactly noon you'd send Bob a signal to tell him whether you'd have time for lunch with him or not. Green means yes, and red means no. Now go ahead and try it."

Alice scratches her head. "You're right. I can't do that. I simply cannot do that. I can only notice whether red or green comes up on its own, because what happens is random. It's what the photon randomly decides, so to speak. I cannot influence whether red or green comes up. I cannot tell the photon what to do. Wow, that's interesting. Could it be that the randomness of the individual quantum event, of the measurement result, keeps entanglement from violating the impossibility of signaling faster than light?"

"Indeed," the professor says, "that is really the case, and it's most amazing. Remember that Einstein attacked quantum mechanics in part for its randomness, that events happen without there being any specific cause for the specific individual results. And now, just this randomness saves entanglement from violating his own theory of relativity. Isn't that really funny?"

Alice and Bob get excited: "That is amazing." "Was Einstein aware of that?"

"We don't know. At least I have not come across any letters or anything in an article or a book of Einstein's that says just that."

Bob exclaims, "I've got it! What these quantum particles are able to do is, they can signal to each other faster than the speed of light. But we cannot make use of this, because we cannot force the quantum particles to carry our signal."

The professor smiles. "This is certainly a way to understand the situation. It also tells us something very deep. Namely, it tells us what is meant by 'signal' in Einstein's thinking. A signal must be something by which we are able to communicate some new information to someone else. If we can't influence what's being sent, then it's not a problem if the stuff travels faster than the speed of light."

LOOPHOLES

After a few moments of thought, Professor Quantinger continues. "Actually, there were three loopholes in the early experiments. You learned a little about two of them already. One is the *communication loophole*. This is the question of whether the two measurement stations are somehow able to communicate with each other and thus make sure that the measurement results correspond to the quantum mechanical predictions. This loophole is excluded by the Innsbruck experiment.

"There is another very important loophole. It has already been pointed out by John Bell that the individual choice of which measurement is performed at both measurement stations must be completely free. This means that it should not be determined by any earlier event. Clearly, such a possibility cannot in principle be excluded definitely, because there might be unknown information that influences both measurement choices. But it is possible to exclude some rather reasonable assumptions about such possible influences.

One such assumption would be to say that the hidden information influencing the settings of the random number generators was created together with the photon pairs at the moment they were emitted by the source. Such an explanation would in principle be possible even for the Innsbruck experiment, because it took the photons a while to travel in the glass fibers from the source to the respective measurement stations. Such an explanation was excluded in a recent experiment done by Thomas Scheidl and other members of the Vienna group. What they did was decide which parameter to measure using a random number generator such that it was located a distance away, and create a random number at the same instant when the photon pair was created in the source. That way no signal starting from the source could have influenced the random number generator. The output of each random number generator was then sent to its measurement station and in this way the specific measurement of the photon was set. So in that experiment, two loopholes, the communication loophole and the so-called freedom of choice loophole, were both closed at the same time.

"But there are still two possible improvements for these kinds of experiments. One is to use human experimentalists who decide at the last instant which polarization of the photon will be measured. Doing such an experiment, we assume that humans have free will. You should know that there is a broad discussion presently among psychologists and people doing brain research whether we really have free will or not. But in any case, in such an experiment we would have to have two experimentalists separated by a very large distance. This is because from neurophysiology, we know that it takes at least one-tenth of a second to make a decision. A tenth of a second corresponds to the time it takes to travel a distance of 30,000 kilometers at the speed of light. So, such an experiment would be most conveniently done using one measurement station on Earth and the other one on the Moon with the source on a satellite in between. Another possibility would be to do such an experiment when people are going to another planet, for example, to Mars. They would have to spend a lot of time going from Earth to Mars, so they would have time to do quantum correlation experiments."

"That sounds really exciting!" Bob interjects. "I would love to be part of such an experiment!"

"Me, too," Alice agrees. "Is the second proposal also as exciting?"

Professor Quantinger replies, smiling, "I personally find the second proposal at least as exciting and interesting as the one we just discussed, but unfortunately, you would not have the opportunity to travel into space to do it. In that proposal, we would use signals from very distant stars that could not have had any connection with each other. An explicit possibility discussed is to have each of the two polarizers be operated through light from a quasar. We would use two quasars at completely opposite sides of the sky, because quasars are among the oldest and farthest distant objects we know, billions of light-years away from us." And the professor points with his arms outstretched to the right and to the left, imagining two completely different locations far away.

"We could certainly assume that the quasars are somehow connected with each other via the big bang. But I consider such an argument to be rather far-fetched. And in principle, as we discussed already, a completely deterministic explanation can never be excluded. I am sure experiments of these kinds will be done someday."

"Recently, under the leadership of Paolo Villoresi and Franco Barbieri at the University of Padua, an international collaborative team was able to perform a first proof-of-principle experiment in that direction. Using a telescope in Matera near Bari in Italy, we sent laser pulses up to the Ajisai satellite. This satellite consists of mirror deflectors. Some of our light was sent back to the telescope on Earth. We could actually detect individual photons arriving back on Earth, having been reflected from the satellite.

Alice twirls a lock of her hair absentmindedly and asks, "Didn't you mention a third loophole, Professor Quantinger?"

"The third loophole," the professor replies, "is what's called the *detection loophole*. In all the experiments so far, only a fraction of the photons have actually been measured, typically about 20 percent, as in your experiment. The measured photons confirm completely the predictions of quantum mechanics. Therefore, the data violate Bell's inequality. But advocates of local realistic theories have made a rather interesting point. They suggest that the following scenario would be possible: maybe all photon pairs made by the source taken together would not violate Bell's inequality. More precisely, the assumption is that all photon pairs would be explainable by a local realistic theory. They explain the measurement results by assuming that for some reason, maybe because of some additional hidden variable, the subset of photons detected is selected such that it violates Bell's inequality."

"That sounds like a rather contrived proposal to me," Alice interjects.

"Well, but the position is at least logically possible, even if it may sound very strange. Why should the world be constructed so that the detectors operate in a way that has us believe the world is not locally realistic when indeed it is? But logically, the position is tenable.

"To conclude our discussion of that loophole, the experimental strategy is clear. One just has to do an experiment in which one detects all, or nearly all, particles. Indeed, it turns out that it is enough to count about three-quarters of all particles. Such an experiment is not possible with photons to date, because the detectors are not good enough. But in 2001, Mary Rowe, David Wineland, and their group at the National

Institute of Standards and Technology (NIST) in Boulder, Colorado, performed just such an experiment with ions, that is, atoms that carry a charge, because for ions we have detectors that are nearly 100 percent efficient.

"In these experiments they used two beryllium ions, which were trapped in a clever arrangement of electromagnetic fields. The advantage of this experiment is that one can detect the states of those ions with high efficiency. As expected, the experiment did also show a violation of Bell's inequality. So the detection loophole is definitely closed."

"So, the story is over," Bob says.

"In principle, I agree and I guess most physicists agree," Professor Quantinger replies, "but we have an interesting situation. There are three experiments. Each of them closed one of the three loopholes definitively. Weihs's experiment closed the communication loophole. That experiment showed that an unknown communication cannot be used to explain the results. The experiment of Scheidl closed the freedom-of-choice loophole and Rowe's experiment closed the detection loophole. So the fact that only a fraction of the photons are measured in other experiments cannot be the reason for observing a violation of Bell's inequality. But—and this is really funny—each experiment leaves at least one other loophole open.

"In the Weihs and Scheidl experiments, the photons were only detected with an efficiency below 50 percent. So it would easily be possible that nature exploited the detection loophole there. In Rowe's experiment, the two ions were sitting very closely next to each other in their trap. There, nature could easily have exploited the communication loophole in order to obtain a violation of Bell's inequalities.

"The fact that none of these experiments closes all three loopholes is a straw that local realists can in principle still cling to. It is extremely improbable, to say the least, that nature is so vicious that it uses in either experiment another loophole. Also, currently there exists no reasonable proposal of a theory that would describe this kind of situation. Nevertheless, for reasons of intellectual clarity and completeness, someday an experiment will be done that closes all three loopholes at the same time, and then the issue will be settled forever."

They sit together for a while, quiet, deep in their own thoughts.

Alice breaks the silence. "This is really mind-boggling. Absolutely fantastic. Physics seems to have something to do with what we can do in the world, whether we can send a signal or not. That is what the speed-of-light limit is about."

Bob muses, "Yes, and the choice we make, what we choose to measure, is what decides which feature becomes reality. So it seems that we humans have a great deal of control over the world. That's amazing. How can that be? How can the world and the laws of physics depend so much on us humans?"

"Well," the professor concludes, "I am more cautious. One has to think very carefully about all these questions. I guess you guys should meet with a philosopher someday."

"Well, actually," Alice says, "we are about to go to the mountains in a couple of weeks with a friend of ours who studies philosophy."

"But he is in his first year," Bob says.

"Well, never mind," the professor says, smiling, "I am sure you'll have something interesting to discuss. In any case, it was really nice to do this project with you both and to discuss these questions. If you ever become interested in these issues again, send me an e-mail or call me up. I wish you all the best in the future."

After these words, Alice and Bob thank the professor for the interesting project, and leave.

IN THE TYROLEAN MOUNTAINS

It is a beautiful summer weekend. Alice and Bob decide to go to the Tyrol. They are joined by their friend Charlie, a philosophy student in his first year. The main ridge of the Alps runs from west to east through the Tyrol, separating it into Southern Tyrol, located in Italy, and Northern Tyrol, a province of Austria. The three take a cable car up a mountain, where a fantastic view opens up before them. They see the snowcapped peaks of the 12,000-foot-high main ridge of the Alps. Between the high mountains, quaint valleys and villages are tucked into meadows that go high up the mountainsides. From the top of the cable car line, they start their hike up a narrow trail. After some time, they reach the peak of a small mountain from where they have a gorgeous view of the main chain of the Alps.

ALICE (*after some time*): Isn't the view beautiful?

BOB: But it's only here when we look at it.

CHARLIE: Nonsense. Of course it's there when we're not looking. Or do you think the mountains vanish at night?

BOB: But how can you prove the mountains are there if nobody's looking? In order to do that, you'd have to look.

CHARLIE: I don't have to. I could just set up an automatic camera to take a picture and look at that later.

ALICE: Yes, you're right. Apparently, it doesn't have to be a human who's doing the looking. But still, someone has to look at the picture of the automatic camera, and that also is an observation. Without making any observation at all, whatever form that may take, we can never make the claim that something is there.

CHARLIE: But now you're just splitting hairs! Does anybody truly believe that the mountains aren't there when nobody's looking?

BOB: You sound just like Einstein. During a discussion, he once asked the Danish physicist Niels Bohr, "Do you really believe the Moon isn't there when nobody's looking?"

CHARLIE: Well, at least I'm in good company, but what was Bohr's answer?

BOB: Bohr challenged Einstein to prove the opposite, which evidently you can't do, because in order to find out if the Moon is there, you have to look up.

CHARLIE: This is really confusing.

ALICE (*laughs*): Welcome to the club. We don't really know either, but it all has to do with a project we've recently been working on for Professor Quantinger.

CHARLIE: Ah, the quantum physicist. How do you expect me to understand any of this? I'm just a poor philosophy student.

BOB: In principle our experiment is simple. Professor Quantinger's graduate student John set up a photon source for us.

CHARLIE: Photon? What's that?

ALICE: Photons are particles of light. Light consists of incredibly small particles that are emitted by a light source.

CHARLIE: Ah, and those particles then go into my eyes.

BOB: Precisely. The source that John has set up for us always emits photons in pairs, twins that are exactly the same.

CHARLIE: Ah yes. When you look at one, it's blue, and its twin sister is blue too.

ALICE: That's one possibility. In our experiment, it's not the color we look at, but another characteristic called polarization, but that's an irrelevant detail.

BOB: Both are always alike when we observe this feature on both photons.

CHARLIE: Well, then it would appear that they were made by the source to be the same. To return to my previous image using colors, in one pair of photons both may be blue, in another pair both may be yellow, and in another red, and so on.

BOB: Those are in fact the experimental observations we've been seeing. The two particles are always identical. To use your image, they have the same color, but their color varies from pair to pair.

ALICE: But the problem is, Charlie, that your explanation doesn't work.

CHARLIE: Well, if they are always found to have the same color, how can my explanation be wrong? They must have had the same color from the beginning, coming from the source. That's it, period.

BOB: Unfortunately, no . . .

CHARLIE: How do you know?

BOB: The proof is a bit complicated, but we'll concentrate on the result. We did an experiment that actually implies that both particles cannot have been created with the same color, but that they only acquired the color when the measurements were taken.

CHARLIE: What do you mean by "acquired the color"? If I look over at, say, those cows, they do not acquire their color when I look at them. Their had their color all along.

ALICE: I fully sympathize with what you are saying, but how can you prove it?

CHARLIE: Well, when I looked before, I saw that they were brown.

ALICE: Well, that's right. But suppose you did not look before? Can you prove in any way that the cows over there were brown before you looked or, for that matter, before anybody looked?

CHARLIE: That's hairsplitting!

BOB: Well, that's what quantum mechanics seems to tell us about some features of quantum systems. We are not allowed to assign properties before an observation.

ALICE: In our case, what it means is that the cow could very well have been brown before the first observation. That would not be a problem. In some cases things are allowed to have their properties before we observe them.

CHARLIE (*demonstratively wiping some imaginary sweat from his forehead*): Thank God.

ALICE: But it gets worse. One can prove that this assumption is sometimes wrong in quantum mechanics, namely, the assumption that objects already have the properties that we observe before we observe them.

CHARLIE: What? The cow was not brown before someone watched it?

BOB: OK. We cannot claim that for the case of the cow, but we can prove it for the case of our two photons.

CHARLIE: Why should I believe you?

ALICE: This is the essence of Bell's theorem, which is named after the Irish physicist John Bell, who discovered it.

CHARLIE: I've heard of Bell's theorem somewhere. But honestly, just because something comes from a famous physicist, I don't have to believe it.

BOB: Very well. This is the essence of science. You don't believe something just because someone famous said it. But many people proved Bell's arguments, and they agree with him. So, couldn't you, just tentatively, also accept what we are telling you? If you want, there is a rather simple paper by Professor Quantinger, written for philosophers, that I can give you in the evening.

CHARLIE: For God's sake, no! I don't want to read a physics paper. I'd rather just believe what you're telling me.

ALICE: So you believe that the two photons don't carry their color before you observe them?

CHARLIE: OK, fine. Just for the sake of argument, I accept that my two photons don't have their color before I observe them. But when I observe them, they have the same color. So there must be some sort of hidden mechanism that, when I watch them, makes them blue, for example. Both photons might have the same mechanism.

ALICE: Yes, very clearly that's the next step. After discarding the model, which was wrong, that the two have the same color from the beginning, the next step is to assume that there is some hidden internal mechanism.

CHARLIE: And this mechanism determines the color each particle has when it's measured, kind of like an inner clockwork.

BOB: Yes, that's also what we assumed. And then Professor Quantinger challenged us to experimentally disprove this hypothesis.

ALICE (*proudly*): And we were indeed able to disprove it.

CHARLIE: Sounds like you'll be winning the Nobel Prize soon.

BOB: Unfortunately, other people already did this experiment.

CHARLIE (*chuckling*): Too bad.

ALICE: But give us a break. We're just first-year students. We're still proud of what we found.

CHARLIE: Now you've made me curious. What exactly did you find? Let's see, the two photons have the same color when we observe them, right?

BOB: Exactly.

CHARLIE: But what is the cause of the fact that they have the same color?

ALICE: So far, we basically know only what the reason *can't* be. It can't be the result of the photons' knowing, in any way, what color they are.

BOB: Right. And they also aren't carrying anything with them that determines what color they are.

CHARLIE: What do you mean, they aren't carrying anything?

BOB: Put it this way. Both particles could be born with lists that they take with them on their journey. And when they reach an instrument that measures them, they could take a closer look at what's being measured, for instance color. And then they could check their lists, find that the color should be blue, and then take on that color. And if both particles have the same lists, they'll always display identical results.

ALICE: That's it. That sounds like a perfect model, right? But this model won't work.

CHARLIE: So wait a minute. You are saying that the photons have no color before being measured. And you are also saying that neither of the two photons knows which color it should show when measured.

BOB: That's how it is. Apparently, it's necessary to make a measurement in order for them to acquire a color.

CHARLIE: I've got it! It must be that the measurement paints the photons!

ALICE: Yes, that sounds like it might be an explanation!

CHARLIE: Easy! The two measurement devices simply are the same in the sense that at a given time they always paint both particles the same color.

ALICE: I have a gut feeling that there's a flaw here somewhere, but I can't figure out what it is.

BOB: I think I've got it. These instructions for the measuring devices— they should be the same for both—are in principle no different from instructions for the particles. They can be summarized as instructions at one location or the other, whether yours, Alice's, or mine, no matter whether these instructions are carried by the particle or the device.

ALICE: Oh yes, and we've eliminated the possibility of local realistic theories in general.

CHARLIE: But there's yet another explanation. What if the two measuring devices were able to tell each other, by sending out some form of energy or exchanging information in some manner, what they are about to do every time? In other words, if they were able to decide together that the next observed pair should be blue, the next yellow, and the next red?

ALICE: That possibility has also been eliminated.

CHARLIE: How come?

BOB: It's simple—by placing the two measuring devices so far apart that a signal from one to the other would take much too long to arrive. After all, every signal can travel no faster than the speed of light.

CHARLIE: But that's unbelievably fast.

ALICE: That doesn't matter. You can build electronic measuring equipment today that can switch back and forth between settings extremely quickly.

CHARLIE: What does that have to do with the question we are discussing?

BOB: Well, you can build devices for measuring photons that can very rapidly decide what they want to measure on the next photon. And that happens so quickly, at the last moment before the photon arrives, that no time is left in which to tell the other device what is being measured.

CHARLIE: And that has already been demonstrated by an experiment?

ALICE: Yes, that was done definitely in an experiment performed not far from here in 1998, in Innsbruck, the capital of the Tyrol.

CHARLIE: And this experiment provided the correct results?

BOB: Yes, so we now know definitely, at least for certain specific quantum situations, that the attributes we observe aren't there before we observe them.

CHARLIE: And that's why Einstein asked Bohr whether he truly believes the Moon isn't there when nobody is looking?

ALICE: Yes, of course. But most interesting, Einstein didn't know all these fine points of modern experiments yet.

CHARLIE: But how did he reach his conclusion then?

BOB: He simply assumed that quantum physics is correct, and quantum physics predicts precisely those results that are now being observed in experiments.

CHARLIE: So how can we comprehend this situation? The mountains are around us. They are not there if nobody looks? That's too much for me.

BOB: Yes, this possibility cannot be disproved by logic, even though I don't believe in it.

ALICE: I don't know if anyone really understands the situation fully. I had the impression that even Professor Quantinger is a bit confused.

CHARLIE: But you must have learned something from all this.

BOB: Yes, there are a number of different conclusions you can draw.

CHARLIE: One is, obviously, that the Moon isn't there when nobody is looking.

ALICE: Maybe. This is the possibility that reality doesn't exist if we're not making observations.

CHARLIE: But that's completely crazy.

BOB: Yet it can't be disproved.

CHARLIE: In terms of pure logic, you're right, but I'd still very much prefer a different solution.

BOB: The second possibility is that something isn't right about our idea that things that are a great distance from one another are actually separate.

CHARLIE: How am I supposed to picture that?

ALICE: Well, that could, for example, mean that there are still unknown types of signals that are much faster than the speed of light.

BOB: But more generally, we could also assume that the way we look at space is incorrect, that, for instance, the peak over there isn't really separated from us by such a large distance.

CHARLIE: But, where does space come from in the first place?

ALICE: And what about time?

BOB: What are space and time?

Our three friends remain seated in silence for some time, looking at the beautiful vistas, until Alice suddenly notices that it is already past three o'clock. They jump up and run down to the mountain station of the cable car, which shuts down at four o'clock. With some effort, they reach the last car just in time. Down in the valley once again, they stroll off toward the village of Alpbach.

BOB: I read somewhere that Erwin Schrödinger is buried in Alpbach.

CHARLIE: Who is Erwin Schrödinger?

ALICE: Who is Mozart?

CHARLIE: You're kidding. Everyone knows Wolfgang Amadeus Mozart. He's one of the greatest composers who ever lived.

ALICE: Yes, and he happened to be Austrian, like Schrödinger.

CHARLIE: I can't know every Austrian who ever lived!

BOB: But Schrödinger was one of the inventors of quantum physics.

CHARLIE: Who cares!

BOB: Well, Schrödinger invented an equation, called the Schrödinger equation, that is probably the most important equation ever written down by any physicist.

CHARLIE: Well, it might be important for you, but why should I worry?

ALICE: That's typical. For a guy like you not to know Mozart would be a huge embarrassment. But you don't care whether you know Schrödinger.

BOB: You should know that the equation invented by Schrödinger is the foundation of most modern high-tech inventions. We would not be able to understand how computers work or how lasers work without knowing the Schrödinger equation. It describes the behavior of microscopic particles, both when they are isolated and when they are inside some solid or some other material, like in computer chips.

CHARLIE: Well, well, I am convinced. You got me interested. Where is Schrödinger's grave?

BOB: The graveyards in all these Tyrolean villages are around the churches. People wanted to be buried close to the Lord. And so is Schrödinger's.

CHARLIE: OK, let's go there! You made me curious.

They walk up to the church, which has a belt of beautiful graves around it, most of them with wrought-iron crosses. When they start to look for Schrödinger's, they cannot find it, so they ask an old lady who points them to a small wrought-iron cross next to the cemetery wall. "Over there!" she says. "Everyone in Alpbach knows about him." Walking up to the grave, they notice that something is unusual (Figure 32).

CHARLIE: There's an equation written on this plaque on the cross.

BOB: Yes, that's the Schrödinger equation.

CHARLIE: Aha. How shall I make any sense out of that? This tridentlike Greek letter, I think it's a phi or a psi . . .

Figure 32. The Schrödinger equation written on the grave of Erwin Schrödinger and of his wife, Annemarie, in Alpbach in Tyrol, Austria.

ALICE: That's a psi. It's the wavefunction.

CHARLIE: And the other letters, like, there. Why is that an i . . . ?

BOB: That is the square root of minus one.

CHARLIE: Oh my God. And what is this h that's been crossed out [$\hbar$]?

BOB: That is an h-bar. Planck's constant divided by two pi, the most fundamental quantity in quantum physics. But don't worry what the symbols mean. In order to understand them, you would have to study physics. And Alice and I are just barely there. Just admire the whole setting.

ALICE: For me, this equation is really beautiful, even if I don't really understand it yet.

CHARLIE: You're right. I have to admit, as strange as these symbols look, the equation is beautiful.

BOB: Yes, there is something appealing in the beauty of mathematical equations. It has to do with the fact that physicists try to describe nature as simply as possible. For example, the Schrödinger equation here uses very few symbols, but describes an incredible wealth of different phenomena.

ALICE: For physicists, simplicity of mathematical expression is one of the clues they have about whether a theory is correct or not.

After a few moments of contemplation, they decide to go back to their quaint bed and breakfast and relax. In the evening, Alice, Bob, and Charlie meet over a pitcher of beer. Alice remarks with a witty smile on her face, "I put this whole story of entanglement down in a little cartoon." And she pulls a piece of paper out of her pocket (Figure 33).

BOB (*surprised*): I didn't know that you had an artistic inclination.

CHARLIE: What does it mean? What do you want to explain that way?

ALICE: Well, in the end, entanglement is about how information is connected between two different players. I've shown this with two books, which are entangled with each other.

CHARLIE: But in the top figure, there is nothing to read.

ALICE: Don't be so impatient. In the first picture, I showed that the texts of the two books are entangled with each other. Neither shows any text. You may say that all possible and conceivable texts are contained in the book. In the middle scene, one of the two observers looks at the book.

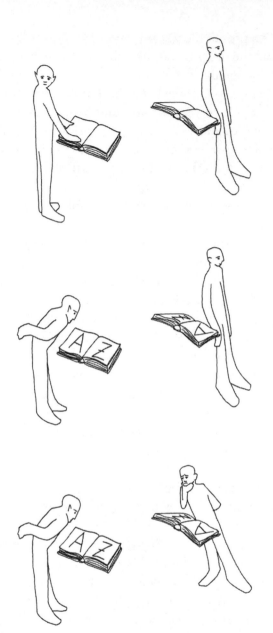

Figure 33. A science-fiction illustration of the entanglement of information in books. The texts of the two books are entangled with each other (*top*) such that neither book carries any text initially. Everything is possible. When one of the books is observed (*center*), it assumes spontaneously and randomly some text out of many possibilities. The other book then instantly assumes the same text. That is the essence of entanglement, which Einstein called "spooky."

BOB: Oh yes! And now the text assumes a certain content, AZ. And the second book also shows the same text. And finally, in the bottom scene, the second observer looks at the book, and they are both surprised that they have the same text in front of them.

CHARLIE: Well, they are surprised the same way that I am surprised about the whole story. That was certainly one of the most important lessons I ever learned.

ALICE: And now you really understand why Einstein did not like the idea of entanglement and why he called it "spooky."

CHARLIE: The grave is kept very nicely and beautifully.

BOB: Yes, that's because Schrödinger's daughter, Ruth Braunizer, still lives in Alpbach. She keeps his study and some of his working documents still in order. This is precious material for future generations of scientists.

ALICE: Guys, I'm getting tired. Let's call it quits for the day.

BOB AND CHARLIE: Great idea.

THE QUANTUM LOTTERY

We've already learned that Einstein criticized quantum mechanics for various reasons. The basic point of his criticism is that some fundamental statements of quantum physics contradicted his own fundamental philosophical beliefs. His first criticism was aimed at the new nature of randomness in quantum physics. He voiced that criticism for the first time in public at the annual meeting of the Gesellschaft Deutscher Naturforscher und Ärzte (Society of German Natural Scientists and Physicians) in 1909 in Salzburg. There, he expressed his "discomfort" about the new role played by chance and randomness. Even at that early point, Einstein recognized that there was no cause for the individual quantum event, not even an unknown one. He saw that for the individual quantum event an explanation of cause and effect was not possible. In a famous letter in 1926 to his fellow physicist Max Born, Einstein wrote, "The theory . . . does not get us closer to the secrets of the old fellow. In any case I am convinced that He"—meaning God—"does not play dice." Today, we have learned to accept that randomness is a fundamental feature of the quantum world. Furthermore, we even use it technically.

How can quantum randomness be put to work?

One area where quantum randomness has been put to work is quantum random-number generators (QRNG). Random-number generators are devices that produce sequences of random numbers. Such sequences are important for many calculational problems in modern computers.

Modern computers generate such sequences using complex algorithms. These produce sequences of numbers that appear to be random, but are not really, precisely because they are the result of calculations.

These are called pseudo-random-number generators. The problem with pseudo-random-number generators is that they repeat themselves after some time. Also, having been started from the same initial state, they generally follow the same sequence of numbers.

Quantum random-number generators provide infinitely better results. Quantum randomness guarantees that no intrinsic mechanism determines which random number appears at any instant. Therefore the sequences of random numbers produced by quantum random-number generators have two distinct advantages. First, they have no internal structure whatsoever, and second, they never repeat.

We have already met this new kind of randomness in some situations. One is when a photon polarized at 45 degrees meets a polarizer oriented, say, vertically. Then, the photon has a fifty-fifty chance of passing through the polarizer. It is completely random, and there exists no reason whatsoever why a specific photon passes through or does not.

Another case where randomness came up was the question of where on an observation screen an individual particle lands after passing through the double-slit assembly. There, we can give only the probability of the particle's landing on some specific spot. But besides that, where an individual particle really lands is completely open.

Furthermore, on a fundamental level, quantum randomness keeps entanglement from violating Einstein's own theory of relativity. The reason is that although the observations on two entangled particles are perfectly correlated, this cannot be used for sending a message faster than the speed of light, because the observers have no control over what the measurement outcome will be, a result of its randomness.

So, while randomness is very important from a fundamental point of view, it can also be put to work in the mundane environment of computers and random-number generators.

A great way to construct a quantum random-number generator makes use of half-silvered mirrors, today generally called semi-reflecting mirrors (Figure 34). As the name says, half of the light that hits the mirror is reflected and half of it passes through. A light beam that hits such a mirror is split into two beams. Therefore, these semi-reflecting mirrors are also called beam splitters. Imagine Alice and Bob standing on opposite sides of a semi-reflecting mirror. Both would be able to see each other through the mirror, together with their own image reflected by the mirror.

Figure 34. Alice, standing in front of a semi-reflecting mirror, would be able to see both herself and Bob, who is standing behind the mirror. Both images occur with half the light intensity.

What happens if a single photon hits a semi-reflecting mirror (Figure 35)? Being a quantum particle, a single photon cannot be split into two half-photons. It must end up either behind the mirror or in front of it. The chances for both happening are the same, so there is a 50 percent probability of the photon passing through and a 50 percent probability of it being reflected. If we place one photon detector behind the mirror to register the photon if it passes through and another detector in front to register the photon if it is reflected, one and only one of the two detectors will register the photon and provide an electric pulse. There is no way to predict which of the two detectors will register it.

This feature of beam splitters is useful for constructing random-number generators. We simply point a stream of single photons at such a beam splitter and place our two detectors at the output. For each photon, both detectors have the same probability of registering. So the two detectors will click in some random sequence. Modern computers all operate on the binary system, which is based on the numbers 0 and 1. So, let us associate a click of the detector behind the beam splitter in the transmitted beam with 0 and a click of the detector in front of the beam splitter in the reflected beam with 1. Then, the stream of photons will produce a random sequence of zeroes and ones. Such a quantum random-number generator was actually built by my student Thomas Jennewein and colleagues when he was a student at the University of Innsbruck a few years ago. A small fraction of the huge sequences of random numbers generated is reprinted here.

```
1111011010100101011011110011001110101111110010110
1100101011000100001001000110000111010010101011101001
1001001100110110101010101000100110001011010111101101
0001100101110000110010100110000101011000111001010 0
1101101110110100001100110101101001111110001001010 1
1010001000010000100101010100000110000010100110110
1011110011001100110101000100111100000000010011000 1
00011001111011100000000010001100110101010101011111
0101100100010110001010010001010 1
```

There are a number of tests we could do to find out whether the random-number generator operates properly, producing good sequences of random numbers. One possibility is simply to check whether we get

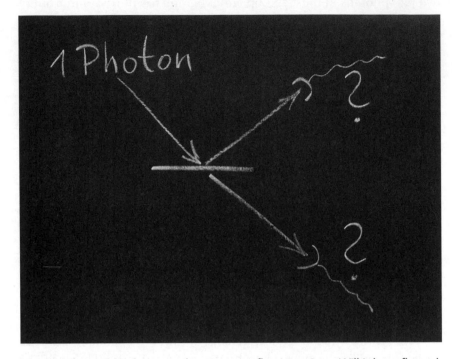

Figure 35. A single photon strikes a semi-reflecting mirror. Will it be reflected, or will it pass through? Which of the two detectors will register the photon? This is an example of irreducible randomness. It is pure chance, without any hidden cause, where the photon will end up.

about the same number of zeroes as ones. Another, more advanced possibility is to look at whether, for example, the sequences 00, 01, 10, and 11 appear equally often. There are many other testing methods, some very sophisticated, including mathematical applications of the random numbers produced.

Jennewein sent a large sequence of his random numbers to specialists who are interested in good random numbers. He asked them to test his sequence in all conceivable ways. Their investigations showed that what he had sent them were the best random-number sequences they had ever tested. That result not only provides strong confirmation of the randomness of individual quantum events, but it also shows that random-number generators based on quantum randomness can be quite useful.

A very general question remains: Is it possible to prove mathematically that a given sequence of numbers, in our case, of zeroes and ones, is really random and has not been generated by some mathematical formula, by some algorithm? The problem is that such a mathematical proof is in principle impossible. There is an important mathematical conjecture stipulating that there are numbers, for example, $\pi = 3.14159265 \ldots$, that contain all possible sequences of digits with each sequence occurring as often as in a random number. Somewhere in the binary representation of π, there is the sequence 0000000000, somewhere else is the sequence 0101010101, and somewhere else is the random-looking sequence 0100101100.

Suppose you throw a die three times. What is the probability of getting a random sequence, like ⚃ ⚀ ⚅? The probability of getting ⚃ on the first throw is 1/6. Likewise for the other two throws. So the probability of getting the specific sequence ⚃ ⚀ ⚅ is $1/6 \times 1/6 \times 1/6 = 1/216$. In other words, on the average, you would have to throw the die 216 times to obtain the specific sequence ⚃ ⚀ ⚅.

What, then, is the chance of getting ⚅ ⚅ ⚅? Well, the probability is exactly the same. The chance of getting ⚅ is 1/6 for each throw. So, the total probability of getting ⚅ ⚅ ⚅ is $1/6 \times 1/6 \times 1/6 = 1/216$! The probability that the random-looking sequence ⚃ ⚀ ⚅ will appear is equal to the probability of getting the sequence ⚅ ⚅ ⚅. Neither sequence is more random than the other, even though ⚃ ⚀ ⚅ looks more random.

Given the mathematical unprovability of the randomness of a specific sequence, we have to rely on our physical knowledge of the inner workings of the random-number generator. A down-to-earth example is the game of roulette. The point of the game is that the roulette ball has the same chance of landing in any of the slots. Casino operators go out of their way to guarantee that their roulette wheels are well-balanced and mechanically perfect in order not to give an edge to some numbers over others.

The huge advantage of a quantum random-number generator is that the individual quantum event is by itself absolutely random. The individual result cannot be considered to be predetermined in any way. There are no physical processes that are not quantum, where randomness has that same quality. Therefore, quantum random-number generators are the best possible ones to rely on.

QUANTUM LOTTERY WITH TWO PHOTONS

We discussed the semi-reflecting mirror and noticed that when Alice stands on one side of the mirror and Bob on the other, they both see both their own reflected image in the mirror and the other person through the mirror. The light going from them to the mirror is split by the semi-reflecting mirror. Half of it goes through the mirror, and half of it is reflected back. Again, things become more complicated and more interesting when we enter the quantum world.

Let us assume two photons strike a beam splitter, one from each side (Figure 36). What will happen? We have learned that each photon has the same chance of being reflected or transmitted. That means that there are just four possibilities.

- Both photons are reflected and thus remain on their own side of the mirror.
- Both photons are transmitted and thus each photon ends up on the other side.
- The photon coming from the upper beam is reflected, the photon coming from the lower beam is transmitted, and so both end up on the upper side.
- The photon coming from the upper beam is transmitted, the photon coming from the lower beam is reflected, and so both end up on the lower side.

Now, all four possibilities occur with equal probability, because the two photons are clearly independent and each one can do whatever it likes. The seemingly logical conclusion is that in half of the cases, we

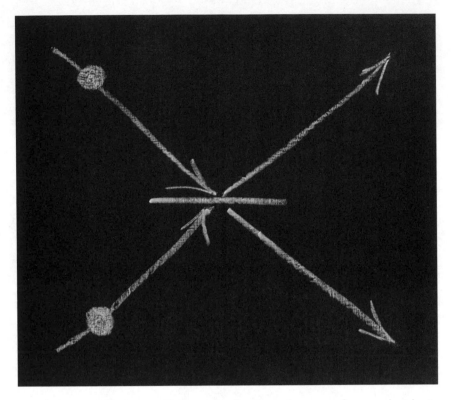

Figure 36. Two photons strike a semi-reflecting mirror, one from each side. In which of the output beams will the two photons end up? The balls are for illustration. They should not be taken literally.

will have one photon in each of the outgoing beams, the first two possibilities discussed above; in one-quarter of the cases, both photons will be in the upper outgoing beam, the third possibility; and in one-quarter of the cases, both photons will be in the lower outgoing beam, the fourth possibility.

Such an experiment was actually done. It was performed in 1987 by Chung Ki Hong, Zhe-Yu Ou, and Leonard Mandel at the University of Rochester. Their experiment disagreed with the simple prediction we just made. The result was that the two photons always ended up on the same side. So there is never one photon in each of the outgoing beams. In half of the cases, both photons will be in the upper outgoing beam, and in half of the cases, in the lower outgoing beam (Figure 37). How can that be understood?

What happens is quantum mechanical superposition, which we actually learned about earlier. Each of the two photons is in the superposition of being in the upper beam or in the lower beam after it has encountered the mirror. Strictly speaking, the statement that the photon is either on the upper side or on the lower side can only be made after an actual measurement is done, that is, after a detector actually registers.

We also learned earlier that superposition occurs when it is not possible, not even in principle, to distinguish which of two (or more) possibilities actually is the case. In our situation, we have two such possibilities. One is the possibility in which both photons are transmitted, and the other is the case where both photons are reflected. If the two photons cannot be distinguished from each other, then by looking at the photons after the beam splitter, we cannot tell which of the two possibilities actually happened. All we have is one photon in each beam, and we have no way of telling where a specific photon came from. Was it from the upper beam, or from the lower beam?

Thus we have to superpose these two possibilities. That is, we have to superpose the possibility that the two photons in the outgoing beam both were reflected with the possibility that the two photons both were transmitted. What is now the result of the superposition? We remember from the double-slit experiment that there are two extreme cases of superposition. In one case, the superposition is called destructive, and in the other one it is called constructive. Now, for our two photons, is the superposition destructive or constructive? We will come back to that in a moment.

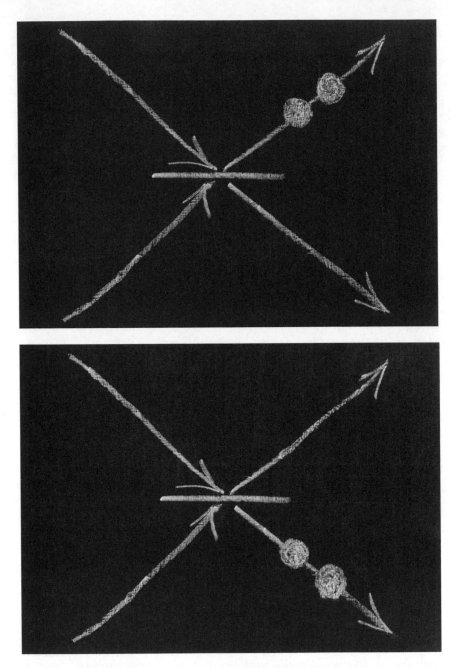

Figure 37. In this experiment, two undistinguishable photons always emerge together on the same side of the semi-reflecting mirror. They are either both in the upper outgoing beam (*top*), or they are both in the lower outgoing beam (*bottom*). Which one they will finally be detected in is completely random.

First, we need to analyze again a fundamental question. In order to apply the idea of superposition here, we have to check whether the two possibilities are really indistinguishable. How could the two photons be distinguished? They could, for example, be distinguished by their wavelength (that is, their color) or by their polarization. Let us assume that both are identical for both photons. They have the same polarization and the same wavelength. That means there is no way by looking at the polarization of the outgoing photons or measuring their wavelength to find out which photon was which. One important way of distinguishing the two is left. They can easily be told apart if they don't arrive at exactly the same time. If the photon coming from the upper beam arrives a little earlier than the photon from the lower beam, then we can simply distinguish them by their time of arrival at the detector.

In their experiment, Hong, Ou, and Mandel carefully eliminated all the possible ways of distinguishing the photons. They made sure that the two photons hit the beam splitter at exactly the same time, with an accuracy of within a few femtoseconds. One femtosecond is 10^{-15} second, a millionth of a billionth of a second. Both photons therefore were completely indistinguishable, and we indeed have to superpose the two possibilities of both photons being reflected and both photons being transmitted. The question now is whether this is destructive or constructive interference. The experimental result gotten by Hong, Ou, and Mandel indeed tells us that these two possibilities extinguish each other, that there is destructive interference. This is because Hong, Ou, and Mandel never saw one photon alone in each outgoing beam. They always came out together (Figure 37). The reason for this is rather intricate.

Would we expect this kind of behavior for all kinds of particles? The answer is no. There are actually two different groups of elementary particles—bosons and fermions. Bosons are named after the Indian physicist Satyendra Nath Bose. Fermions are named after the Italian-American physicist Enrico Fermi. Photons are bosons and electrons are fermions, to give two examples. Bosons like to stick together; fermions like to be separate. And it is the general view that the behavior we just saw is a confirmation that photons are bosons. They like to be together; they like to end up together in the same outgoing beam, in our case. But, as we will later see, this view is too limited. It does not take into account the possibility that the two photons might be entangled with each other. We will discover an important consequence when we examine that possibility, a consequence that is important for quantum tele-portation experiments.

Actually, the experiment done by Hong, Ou, and Mandel had another interesting twist. As we just learned, we obtain this quantum behavior when the two photons are indistinguishable. Hong, Ou, and Mandel therefore made sure that the two photons could arrive at the same time at the beam splitter. But actually, they were able to vary the delay of one photon relative to the other one. So, they also could check for when the two photons did not arrive at the same time, but at slightly different times, at the beam splitter. In that circumstance, the photons *are* distinguishable. What do we expect then? Well, we expect exactly the behavior we discussed above. We expect that in half of the cases, the two photons will come out in different beams, in a quarter of the cases, both will come out in the upper beam, and in a quarter of the cases, both will come out in the lower beam. This is indeed what Hong, Ou, and Mandel saw. Furthermore, they were able to vary continuously this time difference, so they were able to vary continuously the measure of distinguishability. As they increased the time difference, they intro-duced more and more distinguishability, and thus, they saw a slow dis-appearance of interference. They saw a correspondingly slow increase in the cases where the photons arrived in different beams

QUANTUM LOTTERY WITH ENTANGLED PHOTONS

We now might ask a completely new question. What happens if the two photons striking the beam splitter, as in Figure 36, are entangled with each other? Is there anything new we can learn? Let us first recall that the condition for the observation that both photons always arrive together in either outgoing beam, as in Figure 37, was that the two photons are indistinguishable. We considered the case where the two photons carry the same polarization.

Now, suppose that they are entangled in polarization. Entangle-ment means that neither photon carries any polarization before it is being measured. Now, since the measurement happens after the pho-tons leave the beam splitter, how should we apply our rule? Well, again, the crucial point is distinguishability. Let us now assume a specific case of entanglement. This is the one we have already discussed a few times, that entanglement means that both photons, if measured, carry the same polarization. So suppose we measure horizontal H or vertical V polarization. That means if we measure one photon and we find randomly that it is, for example, H-polarized, then the other one will turn out to be H-polarized. We can actually easily imagine a situation where the two photons have the same polarization along whichever direction we might measure them.

It turns out that in our experiment, if two such entangled photons strike the beam splitter, as shown in Figure 36, the two photons always come out together, as in Figure 37. That means if we detect one photon in the upper outgoing beam, the other one will also be there, while if we detect one photon in the lower outgoing beam, the other one will also be in the lower beam. So, the two photons behave as if they had identical properties, even though they do not have them yet, since before the measurement, they do not yet have their individual polarizations. This is quite remarkable. It is not the property the photons carry that decides what they do, but it is the feature that they will exhibit the same polarization if they should be measured later on. So it seems that even two entangled photons behave like two identical photons with the same polarization.

But wait a minute. We have not considered all the possibilities of entanglement. There is also a kind of entanglement where the two entangled photons show different, orthogonal, polarization when they are measured. So again, neither photon carries any polarization before the measurement. But when one is measured, it randomly gives the answer H or V and then the other one shows V or H, the different possibility. And there is actually one possible kind of entanglement where the two photons will always show orthogonal polarization however we measure it! In that kind of state, the photons always prefer to be different, so to speak.

What will now happen if the photons in such a state strike our beam splitter (Figure 36), one from each side? The interesting point now is that the two photons are different in a very novel way. They do not carry polarization yet. But if we should decide to measure them, they will demonstrate that they are different. Suppose we now ask the question again: If we catch one photon in each outgoing beam, will we be able to distinguish where it came from? The first reaction might be that we just have to measure the polarization, and since the polarizations of the two photons are different, we must be able to tell.

But wait another minute. All we know in the beginning is that the two photons will exhibit orthogonal polarization, but we do not know which photon is which. So suppose we catch in the upper outgoing beam a horizontally polarized photon. We do not know which beam it came from, because it could have been from either. The two photons were not polarized. So in other words, we cannot distinguish, if we catch one photon in each of the outgoing beams, whether they both were reflected or whether they both were transmitted. Therefore, we must have interference. The only question is, Is it constructive or destructive interference? For what comes now, there are clear theoretical reasons, which we cannot give here, because that would mean we have to dive deeply into quantum mechanics.

But let us take recourse to experiment. That experiment actually was done for the first time by myself and others in Innsbruck in 1996. It turns out that only the case shown in Figure 38 occurs. There is always one, and only one, photon in each outgoing beam. They never come out together. Remember, the two photons are in that interesting entangled state in which they are always polarized orthogonal to each other; the second photon always exhibits a polarization orthogonal to whatever polarization we measure on the first photon.

How can this behavior be understood in terms of interference? We noted that there are two possibilities for how one photon of each pair may end up in each outgoing beam. They are either both reflected or both transmitted by the beam splitter. Here, these two possibilities interfere constructively. There is no possibility left for the two photons to emerge together in either outgoing beam. So there is no room left for either case shown in Figure 37 to happen.

It turns out that this feature is related to the fact that the two photons will appear to be orthogonal in polarization, whatever we measure. They are always different. They always have different states. This is exactly what fermions do. Remember that we introduced two kinds of particles, bosons and fermions. Bosons like to be the same; fermions always like to be different. And now, most significant, while photons are bosons, they can also be in states where they like to be different. This is exactly the situation we have in the experiment just discussed. The reason why photons can be in such states is that they have more degrees of freedom accessible to them. Besides being different in polarizations, in our experiment they are also in different paths toward the beam splitter.

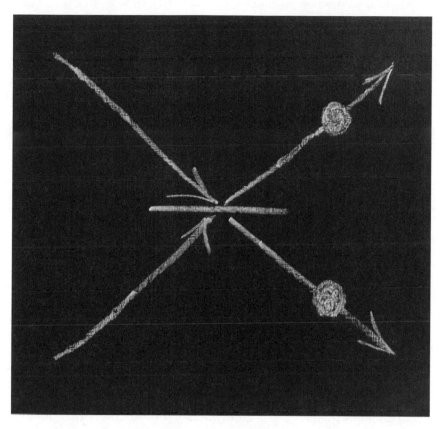

Figure 38. Two photons, which are entangled in such a way that their polarization is always different, orthogonal, whatever way we measure it, strike, one from each side, a beam splitter. In this situation, there will always be just one photon in each outgoing beam. The situations shown in Figure 37 never occur.

So, we have arrived at a very interesting conclusion. If we send two photons separately into a beam splitter, they usually behave well, like good photons, namely as bosons, the way they should. In these cases, they end up together in either of the outgoing beams. So, finding two photons in one of the outgoing beams does not tell us much about their state of polarization. It could be, for example, that the two photons carried the same polarization, say, both H or both V, from the beginning. It could also be that they are entangled in such a way that they will have the same polarization if measured.

But there is this other case where the photons behave completely differently. This is the case where the two photons are entangled such that they are always different in polarization, whatever way we measure the polarization. In that case, they do not like to behave like nice little bosons. In that case, they behave like fermions even if they are bosons; they always end up in separate outgoing beams.

This has enormous experimental consequences. Just take a beam splitter and send two photons in, one from each side. Then, you determine if both come out in the same beam or each emerge separately in the two outgoing beams. If they come out in the same beam, there is not much you can say about the incoming state. They could have carried their own polarization, or they could be entangled. But if they come out in separate beams at exactly the same time, you definitely know that the two photons were entangled in the specific way we just discussed. So this provides us with a simple procedure for identifying one kind of entanglement uniquely. As a consequence, our discovery that photons can sometimes behave like fermions has turned out to be of crucial relevance in a number of experiments, including those involving quantum teleportation.

QUANTUM MONEY—THE END
TO ALL FORGERY

Sometimes, ideas are too early for their times. For example, we might not always have the technology for building what we create in our imagination. This was true for many of the stories of the French writer Jules Verne, whose tales would surely have been dubbed "science fiction," had the term existed at that time. Often, such ideas spark new developments. That's exactly that happened in quantum physics.

In 1970, the young physicist Stephen Wiesner, then at Columbia University, came up with such an idea. Today, forty years later, it has still not found experimental verification. Wiesner invented quantum money. Quantum money has the great feature that it cannot be forged by anyone. It will also not be possible to forge it in the future, unless quantum mechanics at some future date turns out to be wrong in a fundamental way, a very unlikely development.

One would have thought that many institutions, for example the Federal Reserve Bank in the United States, might have jumped immediately at the idea. After all, many millions of dollars of false bank notes turn up all around the world every year. But Wiesner's idea provoked no reaction whatsoever in the business or banking communities.

Even worse, Wiesner was not even successful in trying to publish his idea in a scientific journal. This is a sign of how far ahead of its time it was. It took more than ten years for his paper to finally appear in a journal—not even known in the physics community. It was published by the Special Interest Group on Algorithms and Computation Theory of the Association for Computing Machinery (ACM). Nevertheless, Wiesner's paper was the very first in a new field, the application of fundamental ideas of quantum mechanics to encode and transmit information.

Wiesner's idea is basically quite simple. Every bank note anywhere in the world has a unique number printed on it. These numbers help banks to track the flow of money. They are also useful in some situations, for example, to track money that has been extorted by kidnappers or through blackmail. The numbers on bank notes are clearly visible and can be read by anyone.

Wiesner's idea was to use quantum states for the serial number printed on a bank note (Figure 39). In principle, the idea is correct, although its technical realization still has to see the light of day. One possibility would be to place horizontally or vertically polarized photons somewhere on the bill. One could, say, catch a photon between two super-tiny perfect mirrors, one on the front and one on the back of the bank note. Alternatively, instead of photons, one could use other particles—for example electrons, exploiting their specific property of spin. But in practice, none of the technical realizations have been investigated as they are too demanding for today's quantum technology. Since the physics is the same, we can just use the polarization of photons in analyzing the idea, since we know it so well by now.

It is essential to Wiesner's idea to use not only horizontally and vertically polarized photons, but also photons oriented at 45 degrees rotated to the right or left from vertical. We may call these polarizations S and T, respectively.

So, a typical sequence on the bill might be HSVVSTHSV . . . To forge such a bank note, it would be necessary to read the number in order to print a new bill with the same number. The serial number is indeed necessary for a perfect forgery, since any national bank issues only bank notes with a limited fraction of all possible sequences. Forged notes with numbers outside the set of legal serial numbers can easily be detected.

How would anyone go about reading the quantum number on such a novel bank note?

To read our sequence HSVVSTHSV, our forger has to measure each photon's individual polarization. For the first photon, he gets the answer H. Already the next photon presents him with a problem. If he continues to measure in the H-V basis, he would randomly get either H or V as his result for the second digit. This would not give him any information on whether it was S or T. He would only be able to get the correct result, S, if he knew that the second photon was encoded in such a way as to be rotated by 45 degrees. So, in order to read the quan-

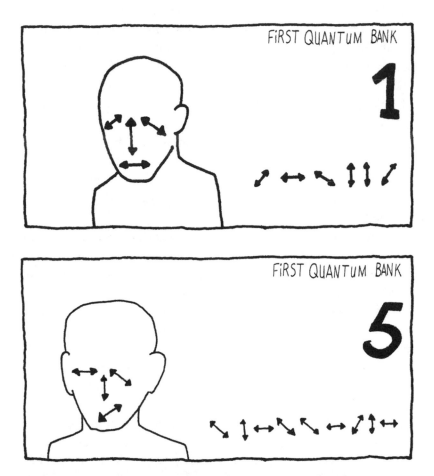

Figure 39. Unforgeable quantum money. A number unique to each quantum bill is printed using quantum bits (*qubits*). Since quanta cannot be cloned, such quantum money cannot be forged. Actually, the specific state of the qubit, indicated by the double arrows, is only shown for illustration. It cannot be seen directly, and it cannot be measured directly unless one already knows what the number is. This information would be restricted to the national bank printing the money.

tum number correctly, we have to know whether each individual photon was encoded in the usual H-V way or in the S-T way. The forger has to know this for each single quantum digit. He has to know the sequence of the orientations of the polarizers. But this is information that only the national bank has. Actually, the national bank keeps this information secret for every quantum bill and uses it in order to identify whether a bill was forged or not.

In Wiesner's idea of unforgeable quantum money, a few fundamental concepts appear for the first time.

The first idea is the encoding of information in two different kinds of orthogonal quantum states, either H-V or S-T. Measuring a quantum system in the wrong basis provides no useful information to the forger whatsoever, only random answers. This method is today called *conjugate coding*.

Another important idea contained for the first time in Wiesner's paper is that quantum states cannot be cloned. This famous *no-cloning theorem* was later mathematically proven by William K. Wootters, then at the University of Texas, and Wojciech Hubert Zurek, then at the California Institute of Technology. The no-cloning theorem says that it is impossible to build a machine into which you can send any arbitrary unknown-quantum-state particle and get two particles to come out in that same state, one the original and the other one a perfect copy. It is the no-cloning theorem that keeps the forger from being able to copy quantum money.

There is actually a possible consequence of the no-cloning theorem in biology. If the hereditary information of living systems were found to be encoded, at least sometimes, in quantum states, then cloning of organisms would be impossible. Today, it is the general understanding in the biological community that the information carried in our DNA is classical information in the sense that it has a well-defined state. But who knows? Maybe someday, someone will discover an exception.

FROM CLASSICAL BITS TO QUANTUM BITS

In his proposal for unforgeable quantum money, Stephen Wiesner unknowingly introduced the idea of what is today called a *qubit*, or *quantum bit*.

All modern digital computers run on bits as the elementary pieces of information. A bit can have one of two states, 0 or 1. A computer contains physical realizations of these bits. Basically, any physical state or physical feature can serve to encode the states 0 and 1. One of the simplest realizations, for example, is a knot in a handkerchief, where the absence of a knot means 0 and the knot in the hanky means 1. Another possible physical representation of a bit would be the position of a switch. "Off" means 0 and "on" means 1. Such switches were actually used in some of the first electrical computers. The switches were electric relays, turned on and off by electrical current running through the computer.

In modern computers, the bit might be realized as a specific voltage in a circuit, or a pit on a CD, or the magnetization of magnetic tape, or whatever. Two features of the physical representation of the bit from a physics point of view are important. First, the two states corresponding to 0 and 1 should be stable and not change into each other on their own. Second, they should be easily identifiable.

The same holds true for communication technologies. Most of today's high-speed communication is done with light. A light beam is modulated and thus imprinted with, for example, speech or a TV program or whatever. An important technological development is to use less and less light to encode a given amount of information. Since light consists of particles, our familiar photons, we might ask what happens if we use fewer and fewer particles of light to encode one bit of information. A limit is clearly reached when one bit is carried by one photon, one quantum of light. To stick to our example, linear polarization might be used to encode information on the photon. Then, H polarization might correspond to 0, and V polarization to 1. Once again, we have two states that are easily identifiable as carriers of information.

Once we use individual quantum particles as carriers of information, completely new phenomena become possible. One of those is quantum superposition. We learned earlier that a photon can be not only in the polarizations H and V, but also in superpositions, for example, polarized at 45 degrees (see Figure 15). This is a superposition of equal parts H and V. In the language of information theory, such a quantum bit, or qubit, can be thought of as being in a superposition of 0 and 1. So, in a sense, it carries both kinds of information at the same time. Only a qubit can be in such a superposition. There is no way a classical bit can achieve that. Consequently, new ways of encoding and transmitting information result. Wiesner exploited this very possibility when he wrote a quantum number onto his bank note. The individual bits of his quantum number were actually qubits.

The notion of such two-state systems has been around in physics for a long time, and systems of two quantum states are among the simplest we can study to learn about the fundamental features of quantum mechanics. In the old days, they were simply called two-state systems. The invention of the name "qubit" by Ben Schumacher of Kenyon College in 1993 was actually something like coining a brand name for the whole quantum information community. For example, the home page of the Centre for Quantum Computation at Oxford University is at www.qubit.org.

Anyway, Wiesner's idea of unforgeable quantum money was followed later by a number of proposals using quantum states to transmit or process information. Of these, we already discussed quantum cryptography, first proposed by Charles Bennett and Gilles Brassard in 1984 using single qubits, and in 1991 by Artur Ekert of the University of Oxford using entangled states.

That qubits can also be entangled with each other is another feature of quantum physics that goes beyond classical physics. A rather interesting application of entangled qubits is hyperdense coding, proposed theoretically by Bennett and Wiesner in 1992. We will discuss that now.

A QUANTUM TRUCK CAN TRANSPORT MORE THAN IT CAN CARRY

What a strange title for a chapter! Well, let's see what it means. Every truck has some load-carrying capacity, say, one ton. When you put more in the truck, it is overloaded, and something, say, its axle, might break down. If we want, we can consider the qubit to be the tiniest truck possible. What it carries is information. We will see in this chapter that a single qubit can indeed be used to send one bit of information, so this is its information-carrying capacity. But most interestingly, if the qubit is part of an entangled pair, it can actually be used to send two bits of information, which is more than its information-carrying capacity.

Let's first address a simple question. If we say that information is represented by bits whose value is 0 or 1, what do we really mean? What sort of information is 0 or 1?

This question by itself does not make much sense. We need to talk about what information means. There is no general agreement about the meaning of information. However, we might say that our knowledge about the world is expressed in statements.

One such statement could be "It is raining."

Another somewhat more complicated possible statement is "The number of tall pairs of twins with blue eyes cannot be larger than the number of tall pairs of twins with brunet hair plus the number of blond pairs of twins with blue eyes."

Any such statement is either true or false. Let's ignore situations where it's hard to decide: sometimes, one is not quite sure whether it's raining or not. There are borderline cases. In other situations, we might in principle not even be able to clearly answer a question like "How many angels can sit on top of a needle?" But for simplicity's sake, let's now consider only those cases where it can clearly be said whether a statement is false or true. Some examples:

It is raining.	False, at least at the moment I am writing this down.
The dollar is the currency of the United States.	True.
This book has fewer than ten pages.	False.
The number of tall pairs of twins with blue eyes cannot be larger than the number of tall pairs of twins with brunet hair plus the number of blond pairs of twins with blue eyes.	True, except for entangled quantum twins.

So, there are only two possibilities, true or false. Our bit also only has two possibilities, 0 or 1. So we can easily represent the answers true and false as bit values. All we have to figure out now is whether 0 means "true" or "false." Then 1 will mean the other. Now, there is not much to figure out here. It's a matter of agreement. If we want to talk to each other, sending each other bits, we need to agree what 0 and 1 mean. It is the customary, but by no means the necessary, choice to have the following identification:

True: 1
False: 0

Likewise, we can do the same with yes and no. We can have 0 represent "no" and 1 represent "yes."

If we now wish to communicate with someone, we have to choose an information carrier. Bob, the sender, wants to send something to Alice. This something has to be represented by the 0 or the 1 value of the bit. Alice and Bob may choose individual particles of light, our photons. They decide that a horizontally polarized photon means 0 and a vertically polarized photon means 1.

So, Alice asks Bob, "Will you have lunch with me today?"

If she receives a vertically polarized pulse of light from Bob, she is happy. Bob just told her yes. If the light pulse she receives is horizontally polarized, she has to look for someone else to have lunch with.

So Bob can encode one bit of information into his photon. This is the information-carrying capacity of the photon when polarization is used. We might ask whether by using superposition, Bob could send more information. This is not the case. For example, if he sends a photon polarized at 45 degrees, then Alice has only one way of getting the correct answer, namely, to measure it again at 45 degrees. If she measures it in any other way, she gets some randomness in the answer.

So it might seem as if the quantum nature of a qubit does not help, but Bennett and Wiesner suggest a most interesting situation arises if one considers entanglement. The idea of hyperdense coding is actually rather simple in principle, as we can see from the following experiment (Figure 40).

Alice and Bob start their experiment by producing pairs of entangled qubits. In the actual experiment, these are photons. Bob gets one of each pair, and Alice gets the other one.

Bob now wants to encode information onto his qubit. He does this by rotating its polarization. As we have learned, an entangled individual particle or qubit by itself does not have its own proper quantum state, so it does not carry any information. Actually, if we measure a maximally entangled qubit in any basis, we will always get a random answer. So, if we always get a random answer, how can we encode information? The point is that while the qubit does not enjoy its own quantum state, its relation to the other qubit is clearly defined. Together, they form a unique entangled state. So, if Bob changes his qubit, he can actually change the joint entangled state of the two qubits that he and Alice have so that another entangled state results.

Actually—and this is very interesting—just by playing with his qubit, Bob can achieve any of four different entangled states. When he does nothing, the original form of entanglement remains. He can achieve the other three entangled states basically by rotating his qubit in the proper way about three possible orthogonal axes: right-left, front-back, plus-minus. So, Bob manipulates his qubit in one of the four ways and sends it to Alice. Alice then determines in which of the four entangled states both her qubit and the one she just received back from Bob now are.

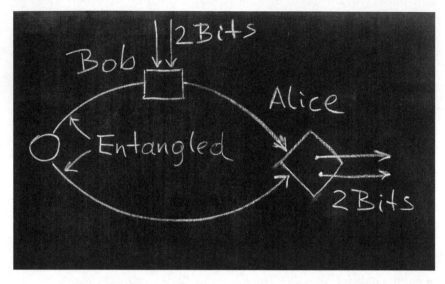

Figure 40. Principle of hyperdense quantum coding. The source (*left*) emits an entangled pair of photons, which travel along two different routes to Alice (*right*). Bob has access to only one of the two photons. But still, he can send two bits of information to Alice. By manipulating his photon, he is able to modify the specific kind of entanglement between the two photons. Alice receives both photons, and thus can extract both bits of information.

And this is the really exciting point. Alice can now, by identifying the entangled state of the two qubits, identify four possible messages—twice as many as if Bob had sent her just one unentangled qubit—since one qubit can carry two possibilities: 0 or 1; yes or no; and so on.

The four possible answers actually correspond to two classical bits of information, because if we consider two bits, each one with the states 0 and 1, we have four possibilities: 00, 01, 10, and 11. These are the four possibilities. Clearly, in the procedure we need two qubits to transmit the two bits of information, but the interesting part is that Bob has to manipulate only one of the two. So, while one qubit still carries only one bit, it can transmit more information than one bit if it is entangled with another qubit.

Bennett and Wiesner's theoretical idea was demonstrated in a real experiment in 1995 by Klaus Mattle, Harald Weinfurter, Paul G. Kwiat, and myself at the University of Innsbruck. In the experiment, the entangled qubits were actually produced as entangled pairs of photons.

These photon pairs were entangled in polarization. Bob transferred the information he wanted to send to his photon. He did that by either doing nothing, that is, leaving the photon unchanged, or rotating the photon's polarization in two different ways. He then sent his photon on to Alice. She also received the twin entangled photon directly from the source. By measuring them together, she was able to determine which way they were entangled. For technical reasons, in that experiment, Alice was only able to distinguish three different possibilities.

So in the end, Alice was able to identify three different messages. This is already significantly more than the two messages that can be transmitted without using entanglement. The experiment provided clear confirmation of the concept. Therefore, an entangled photon can transmit more information than it can carry. A single photon can carry only one bit of information in its polarization, but with an entangled photon, one can transmit more than one bit. This experiment actually constitutes the first application ever of entanglement in a quantum information protocol, confirming that using entanglement opens up completely new possibilities for communication and computation.

ATOMIC SOURCES OF ENTANGLEMENT AND EARLY EXPERIMENTS

How do particles get entangled? How are entangled photons created?

The first method for creating entangled photons was to use atoms. When an atom is excited, it emits light. That is a phenomenon everyone can see by looking at a fluorescent lamp. Important for us is the fact that there are atoms that, if they are excited in the right way, emit two photons right after each other. And in some cases, the two photons emitted are entangled in polarization, just the way we discussed before. In real experiments, the excitation of the atoms is usually done by shining laser light onto them. This kind of atomic source was used in the first experiments that showed a violation of Bell's inequality.

The very first experiment that showed that Bell's inequality was violated and that therefore nature cannot be understood in a "reasonable" way, namely, in a local realistic way, was performed in 1972 by Stuart J. Freedman and John F. Clauser. Clauser was a post-doc at the University of California at Berkeley who collaborated with the grad student Freedman. The experiment has an interesting prehistory.

Bell's 1964 paper did not get much attention from physicists in the beginning. One exception was Abner Shimony. He immediately realized that this was an important paper. Shimony is one of those rare, lucky cases in the history of science of someone who is so well educated in two fields that he can actually combine the two backgrounds in a fruitful, novel way. Shimony was trained as a physicist by the Nobel laureate Eugene Wigner and as a philosopher by Rudolf Carnap. Carnap was a member of the Vienna Circle, a philosophical group that completely changed philosophy in the beginning of the twentieth century. Because of the Nazis, Carnap had emigrated to the United States, where he became a professor at the University of Chicago.

Shimony, then at Boston University, realized that Bell's paper was very important. When a young student, Michael A. Horne, asked him for an interesting topic for a Ph.D. dissertation, Shimony showed him Bell's paper and suggested finding a way to turn this into a real experiment. Horne and Shimony indeed discovered such an experimental possibility, utilizing pairs of photons emitted by atoms.

They took their proposed experiment to the Harvard professor Frank Pipkin, whose student Frank Holt joined them in working out the details for doing the experiment at Harvard. While Horne, Shimony, and Holt worked on the details, Shimony ran into an abstract for a talk submitted by the young Columbia graduate student Clauser to a meeting of the American Physical Society, who made essentially the same proposal.

When something like that happens, scientists are faced with a dilemma. Should they compete, or should they collaborate? In this case, the two parties decided to partly collaborate and partly compete. Clauser joined Horne, Shimony, and Holt in publishing the proposal. Holt and Pipkin then started setting up the experiment at Harvard, and Clauser moved on to Berkeley to independently set up the experiment there using a different atom.

The results of Clauser and Freedman's experiment, published in 1972, clearly showed that Bell's inequality was violated. The world is nonlocal, concluded most physicists. But that, as we discussed already, is not the only possible interpretation.

It is quite remarkable, by the way, that Clauser expected his experiments would show the opposite. He expected that Bell's inequality would not be violated. He did not consider it possible at all that the world could be so crazy that local realism could be wrong. It is actually the sign of an excellent experimentalist that he is able to discover something unexpected in the laboratory. Clauser did not only *not* expect the results he got; he expected exactly the opposite to happen!

But meanwhile, back at Harvard, Holt and Pipkin's experiment did not violate Bell's inequality! They decided not to publish, and instead embarked on a long and inconclusive search for the source of some systematic error. A few years later, Clauser repeated the experiment with the Holt-Pipkin atom and obtained a violation, thus confirming the predictions of quantum physics.

Following the first Freedman and Clauser experiment, there was a series of other experimental approaches, where violations of Bell's inequality were shown with increasing precision, with more details, and where many new aspects came in. A significant step was a series of beautiful experiments by the group around Alain Aspect, in Orsay near Paris. His main collaborators at that time were Philippe Grangier, Jean Dalibard, and Gerard Roger.

At the time Aspect first thought of performing these experiments, he had some discussions with Bell about whether he should do them. Bell's first question was "Do you have a permanent position?" Only after Aspect answered affirmatively did John Bell encourage him to do the experiment. While Bell's reaction was somewhat motivated by the fact that these experiments were extremely challenging and it would therefore take a lot of time to do them, he worried more about the attitude of the community of physicists at that time. Work on the foundations of quantum physics was actually considered not to be quite the right thing to do, an experience the author of this book also had in his career. Luckily enough, the situation has changed.

Aspect's experiments contain three important points. First, he was able to demonstrate a violation of Bell's inequality with much higher experimental precision. Second, he was the first to use polarizing beam splitters, that is, he could use both polarizations of both photons. In the earlier experiments, people used only simple polarization filters, with which one can observe only the transmitted polarizations. Third, he and his group performed the first experiment in which the polarization measurement direction is changed while the photons are in flight.

As before, the question is whether the measurement stations at least in principle could know from each other—or whether the source could know—which specific measurement will be performed on each photon. If that were in principle possible, then some unknown communication could exist, guaranteeing that the results predicted by quantum mechanics actually occur in experiment. That would be a local realistic interpretation, and intuitively acceptable. Certainly, that is a rather improbable possibility, but in principle, it cannot be excluded on logical grounds alone.

Aspect in his experiment switched the measurement of each photon back and forth periodically between two possible polarization measurement stations with differently oriented polarizers. That was done fast enough so that the switch occurred while the photons were in flight.

Aspect's experiment was the first step in the direction of excluding unknown communications. But which unknown communications did it exclude? First we note that the switching was periodical. Thus, which polarization orientation will be active at which time was predetermined. This is not significant in a model where the apparatus has no memory. When we now ask which specific speeds were excluded, we note that the distances between the source and the polarizers were such that communication with the speed of light was not excluded. Nevertheless, the experiment was very important because besides being a real experimental tour de force, it was the first to test the communication loophole question.

THE SUPER-SOURCE AND CLOSING THE COMMUNICATION LOOPHOLE

Sources for entangled photons based on atoms, used in the early experiments, have one big disadvantage. In general, a photon can be emitted by an excited atom into many different directions. Thus, if we happen to catch one photon of an entangled pair, it is by no means certain that we will also catch its twin photon. If we want to detect them at specific polarization measurement stations, we find that many photon pairs are lost.

The best source for polarization-entangled photon pairs today is a very special process inside crystals called *spontaneous parametric down-conversion* (SPDC). Let's not worry about the details but just look at what it does. We take a special kind of crystal and direct a strong laser beam onto it. A photon from the strong beam may then be converted into two photons coming out of the crystal. This looks very much like the decay of one photon into two. For this process to occur, the crystal has to have certain properties. The best crystals for the procedure are artificially made.

For this conversion process, certain laws apply that, for large enough crystals, essentially relate to energy and momentum conservation. This means that the energies of the two new photons must add up to the energy of the original photon from the strong laser beam, and likewise the momenta. But otherwise, the energy and momentum of each photon is unspecified. Therefore, neither of the two photons has a well-defined energy or a well-defined momentum. Thus, the two photons are entangled with each other both in energy and momentum. If one photon is measured, it instantly assumes some energy and momentum. The other photon then has to have the corresponding energy and momentum to fulfill the energy conservation laws.

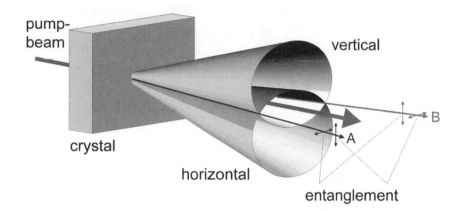

Figure 41. The process of parametric down-conversion to create polarization-entangled photons. A strong laser beam, called a pump beam, strikes a special kind of crystal. Inside the crystal, a photon from the pump-beam may be transformed into two new photons. The photons are emitted along cones with one photon in one cone and the other photon in the other cone. Two photons of a pair are always found opposite each other in the two cones. Within each cone, the photons are polarized. One is horizontal, the other one vertical. The arrangement may be such that the cones intersect along two lines. Photons of a pair emitted along these lines do not know which polarization they carry, but they know that their polarizations have to be orthogonal to each other. Thus, entanglement results.

A more useful source produces polarization-entangled photons (Figure 41). The process again is spontaneous parametric down-conversion, but now of a slightly different kind. Either of the two photons can come out on its own cone (see Figure 41). Furthermore, within each cone, the photon is polarized. In one cone, the photon is V-polarized; in the other one it is H-polarized. Most significant, these cones can intersect. Along the line of intersection, either photon can be either horizontally or vertically polarized and the other one has to be the opposite, but it is not defined which is which. So the states of the two photons can be either HV or VH, but it is completely undecided which of the two situations is actually the case. So, if the arrangement is such that the two are indistinguishable, a polarization-entangled pair emerges.

This source has many advantages. First, its power can be changed by adjusting the power of the incoming beam. Second, the polarization entanglement is of extremely high purity. That means, if we find one photon to be polarized, say, vertically, the other photon will with nearly 100 percent certainty turn out to be horizontally polarized, and vice versa. And third, the two photons come out along very clearly defined directions, so they can easily be used in complicated pieces of apparatus, sent by mirrors around corners, coupled into glass fibers, and so on. This kind of source has become *the* workhorse for many entanglement experiments.

An experiment where the high quality of that source came in very handy was one where the communication loophole was definitely closed. This was done in 1997 by Gregor Weihs, Thomas Jennewein, Christoph Simon, Harald Weinfurter, and myself. We demonstrated in our experiment at the University of Innsbruck that any communication between the two detector stations really can be excluded. As mentioned on page 151, the point there was to use statistically switched polarizers with no periodicity at all. The photon pair was generated in a building in the center of the campus, and the two photons were then guided to two different measurement stations that were finally about 300 meters apart from each other. The important point of that experiment was that the polarizations on each side were switched at the last instance. An electro-optic modulator rotates the polarization in proportion to a voltage applied to it. These electro-optical modulators were driven by quantum random-number generators. The switching was so fast that the polarization orientation was only decided within the last few meters before the photons arrived at the respective polarizers. There was no information present beforehand anywhere about the future orientation with which a specific photon would be measured. Any communication that might have happened telling the photon on one side which polarization is being measured on the other would have had to travel much faster than the speed of light. Since this is excluded by Einstein's special theory of relativity, we can safely conclude from the experiment that communication between the two detector stations cannot be used to explain the violation of Bell's inequality observed in the experiment. In any case, whatever the interpretation, experiment clearly confirms the existence of quantum entanglement even with completely independent measurement stations.

QUANTUM TELEPORTATION AT THE RIVER DANUBE

We are now ready to return to Rupert's teleportation experiment at the river Danube in Vienna, which we learned about at the beginning of this book. We now know all the ingredients for that experiment and we can put them all together.

It is now May, and we drive again down to the island in the Danube where Rupert's laboratory is located. May is beautiful in Vienna. We drive through a lane lined with beautiful chestnut trees. The trees are in full bloom. Again, we walk down with Rupert to his underground laboratory, which is filled with all his lasers and optical equipment. This time, we hope to understand what is going on. We have learned what the essential ingredients of the teleportation experiments are, and here they are in their full glory in front of us. Rupert, obviously proud of his experiment, points at a poster showing its various components (Figure 42). Significant parts of the experiment, he tells us, actually happen inside glass fibers. At first, he shows us the source. There is a huge laser system. It is quite expensive.

Rupert smiles proudly. "You can get a house for a small family for that money.

"An essential point," he tells us, "is that the laser system does not produce a continuous beam, but pulses of laser light in very rapid succession. Each pulse is about 150 femtoseconds long and the system produces about 80 million light pulses every second." We recall that one femtosecond is a millionth of a billionth of a second (1 femtosecond = 10^{-15} second). "So," he says, "while the succession time between two pulses sounds very short, the time separation between them actually is about 100,000 times larger than the width of an individual pulse. It is

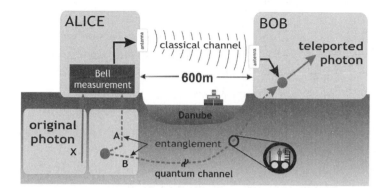

Figure 42. The quantum teleportation experiment at the river Danube. Alice and Bob are connected by two information channels. The classical channel is a radio-wave connection above the river. The quantum channel consists of an entangled pair of photons, A and B, that fly through glass fibers. The glass-fiber cable passes through an underground tunnel under the Danube. Alice teleports the state of the original photon X. For that purpose, she performs a Bell-state measurement jointly at X and A. That way, the two become entangled with each other. Using the classical channel, Alice communicates the specific kind of entanglement to Bob. In one case of the possible entanglement results, Bob's photon is immediately in the state of the original photon X. He does not have to do anything at all. In other cases, Bob has to rotate the polarization of his photon B depending on the information he receives over the classical channel. In both cases, the teleported photon emerges identical to the original photon A, which has lost its own private properties in the process.

like having a lighthouse beacon flash once per day for just one second." He continues by explaining the part of the setup where the photons are created. "That would be a very bad lighthouse beacon, but any seaman knows that the more rarely a beacon flashes, the more important it is. That we have about 80 million flashes per second gives you an idea how short an individual pulse of light produced by the laser is."

Rupert, seeing our questioning faces, explains in more detail.

"You might wonder why we are producing such short pulses of light. This has nothing to do with the importance of the flash, but with quantum indistinguishability. You will see immediately." And he continues explaining the setup (Figure 43).

"Let's envisage an individual light pulse now. It comes out of the laser and then passes through this tiny crystal here. This special crystal produces the entangled photons. It is only 2 millimeters (about 3/52 inch) thick, but what's happening here is most important for our experiment. Two photons fly off at certain angles. Both are entangled with each other. We send these photons into glass fibers using small lenses in front of the fibers, and then on to Alice and Bob. What we just produced was an entangled twin pair consisting of photons A and B. That is the quantum channel that will be used by Alice and Bob to teleport another photon.

"One of the twin photons goes over to Alice close by here on the table, and the other twin photon crosses the river to Bob. That will eventually be the teleported one. But let's get back to our pulse. After having left the crystal, the light pulse goes on and hits this very small mirror."

Rupert points at a small cube of glass standing on its side. "This does not quite look like a mirror; it looks transparent. Visible light is actually not reflected. But this cube has a special coating that is able to reflect the light of the pulses very efficiently. That light is in the UV, or ultraviolet, range, invisible to our eyes. The reflected laser light pulse passes again through the crystal, creating another pair of photons. These photons in principle are also entangled, but we don't use the entanglement this time. One of the two photons passes through a polarizer, which can be adjusted in any way. We can imprint on it any polarization we want. Since it now carries a specific polarization, that photon is not entangled anymore. Finally, it is also put into a glass fiber and sent to Alice. This is the photon whose state is going to be teleported by her.

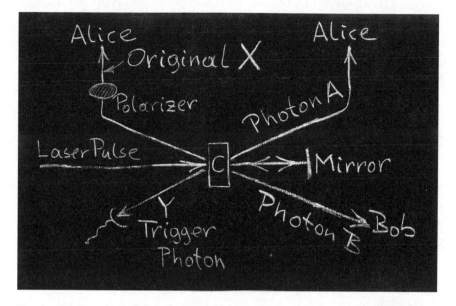

Figure 43. Creation of the photons in the quantum teleportation experiment. A very short laser pulse passes through a crystal C and creates a pair of photons, A and B, which are entangled with each other. These two photons form the quantum channel for teleportation. Alice gets photon A, and photon B is sent over to Bob. The laser pulse is reflected back to the crystal by a mirror. At its second passage through the crystal, it creates another pair of photons, X and Y. The trigger photon Y tells us that the other photon (X) of that pair has been produced. Photon X is the one that is ready to be teleported. It passes through a polarizer that can imprint different kinds of polarizations on that photon. The original photon X is handed over to Alice in order for her to teleport it.

We call it the original X. The other photon of this second pair serves as a trigger. The purpose of a trigger is very simple. If that trigger photon is registered, we know that the other photon of the pair is on its way. This other photon is the teleportee photon X." Rupert smiles.

"So let's recapitulate," he says. "One photon, photon Y, has been registered by the trigger, and we have three photons inside glass fibers. Photon B is on its way to Bob's station on the other side of the river, and the other two photons, photon A and photon X, are heading to a meeting point at Alice's station. Photons A and X meet in an in-fiber beam splitter [Figure 44]. The in-fiber beam splitter, called a fiber coupler, works this way: You have two glass fibers that run along side by side. The cores of the two glass fibers are very close together, such that some of the light can go across from one fiber into the next. And if this is done right, this fiber coupler acts just like a fifty-fifty beam splitter. So, for any of the two inputs, half of the light ends up in one output and half in the other output. We use fibers because the experiment is much more stable that way. In principle, we could have built the thing with mirrors, beam splitters, and so on in free space, but the in-fiber solution is much more stable.

"The two fibers leaving the fiber coupler," Rupert continues, pointing again, "now go to one polarizer each to identify the entangled states of the two photons A and X. Actually, we can identify two such entangled states, also called Bell states."

Rupert now explains that the timing of the experiment is crucial. "One reason for using the short pulses is that we want the two photons—the teleportee photon X and Alice's twin A of the entangled pair—to arrive at the beam splitter at exactly the same time. Now, if they are created within 100 or 200 femtoseconds, we can make them arrive simultaneously at the beam splitter if all the travel times add up precisely. We must therefore make sure that the photon created on the first passage, photon A, travels a somewhat larger distance to the in-fiber beam splitter than the other photon (X) does, because the other photon is created later, during the second passage of our pulse, and we want them to arrive at the same time at the in-fiber beam splitter. The really hard part of the experiment was to adjust all these beam path lengths exactly so that they have the same length within something of the order of 50 micrometers. Initially, that was very difficult, but by now, we have learned how to do it," Rupert announces with a big smile.

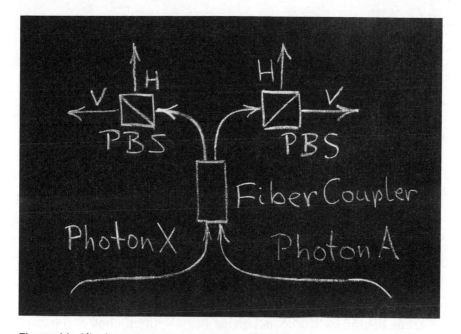

Figure 44. Alice's measurement station in the Danube experiment. Photon X is in the state to be teleported. The point of the whole procedure is to entangle this photon with photon A, which is entangled with photon B, which is on its way to Bob. All these photons are inside glass fibers. The fiber coupler operates just like a beam splitter or a semi-reflecting mirror. Each of the two incoming photons has the same probability of ending up in one of the two outgoing fibers. By measuring these photons behind two polarizing beam splitters (PBS), we are able to project the original photons A and X into certain entangled states. The experiment itself can distinguish two different entangled states. In one of the two entangled states, the two photons end up in the same outgoing glass fiber, either the right-hand or the left-hand one. The entangled state is then identified by measuring one photon each in the H and V outputs of the same PBS. The other entangled state is identified by the fact that one photon is found in each of the outgoing glass fibers behind the fiber coupler. This state is easily confirmed if we find a photon in either of the outgoing beams behind the PBS on one side together with a photon of opposite polarization in an outgoing beam of the other PBS. The fundamental reason that photon A and photon X become entangled through the procedure is basically simple. There is no way to tell which of the photons in an outgoing fiber behind the fiber coupler came from which of the incoming fibers. Thus, they lose their individual identity.

"The point is," Rupert says, beaming proudly, "that we have ultra-fast electronics here. All the detectors work on a time scale of a few nanoseconds. A nanosecond is a billionth of a second. And there are some electronics that identify the Bell state." He points again at the sketch of Alice's measurement station (Figure 44) and continues. "Each of the two outputs of the fiber coupler has a polarizing beam splitter, or PBS, where the photon takes one path if it is horizontally polarized and the other path if it is vertically polarized. In the entangled states that we identify, the two photons are always differently polarized, so one is always horizontally polarized and the other always vertically polarized. For a certain state—physicists call it the antisymmetric, or fermionic, state; I like to call it maverick—it happens that the two photons always go separate ways after the beam splitter [compare Figure 38]. For the other state that we observe, the two photons go the same way, both in either one or the other behind the splitter [compare Figure 37]. Therefore, the logic is simple. If two different detectors (one H, the other one V), one behind each PBS polarizing beam splitter, click at the same time, we know we have the maverick state. If two detectors on the same side behind the same polarizing beam splitter register a photon together, we have the other state.

"It's the purpose of our electronics to identify which of the detectors click together. This information, which of the two states we obtained, has to be passed over to Bob. That's done by a microwave radio link." Rupert points at the cable going up from his assembly and continues to explain.

"The antenna is on the rooftop of our building. Bob will then receive the message and set up his apparatus so that the photon arriving at his place will be turned to the right polarization. Let's walk up to the antenna," Rupert suggests. We take the elevator up to the top level of the building and then climb a narrow ladder to the rooftop.

The antenna points across the dark river. Rupert tells us that there is a similar antenna on Bob's side.

While we get back down again, Rupert continues: "Still, how can we make sure that the microwave signal arrives at Bob's station before the photon does? There are two things that help us. First, the speed of light inside a glass fiber is only about 200,000 kilometers per second, whereas the speed of light in air, and therefore also the speed of our microwave radio signal, is about 300,000 kilometers per second. So, as the separation between Alice and Bob is about 600 meters, the radio signal takes about two microseconds and the photon inside the glass fiber takes about three microseconds. So we have gained one microsecond in time, which in principle should be fast enough to run all our electronics. But just to make sure"—and he smiles—"and because I had extra cable, I coiled the glass fiber around here under the table a couple of times, adding another 200 meters to the total length. This delays Bob's photon by another microsecond, so we have ample time, namely, two microseconds altogether, for our electronics to do their thing. But let's go over to the other side and look at Bob's setup."

This time, we drive across the river to the other side. Bob's setup is actually quite simple.

"No need to climb to the roof," Rupert says. "The antenna looks just the same as the one on Alice's side."

Rupert shows us the cable coming down from the antenna and the glass fiber cable coming out of the wall. The setup consists of a computer and a small optical table.

"This is our breadboard," Rupert says and smiles. "Funny name, isn't it? Nothing to do with bread. But I guess it's a breadwinner for me."

Compared with the optical table on Alice's side, the breadboard looks almost empty. Only very few optical components are attached to it.

Rupert explains, "Here is the cable that comes from the antenna. That's the classical channel providing us with the information about which entangled state Alice obtains in her measurement. If it happens to be the maverick state, Bob's photon B is already in the state we want it to be in, namely, the original state of the teleportee photon X. We have achieved teleportation. If the photon is in the other entangled state, we have to perform a polarization rotation on the arriving photon B in order to change its state to that of the original X. For this we again use an electro-optic modulator [Figure 45]. If you apply a voltage to the crystal, the polarization of an incoming photon is rotated by some angle, depending on the voltage.

"In our case," he continues, "zero voltage is applied to the modulator if the message we receive from Alice via the radio signal tells us that she identified the maverick state. If we receive the message that it was the other entangled state, we apply about 2,400 volts. This rotates the polarization of Bob's photon B so that it's in the state the original teleportee photon X was at the beginning." Rupert points at another PBS:

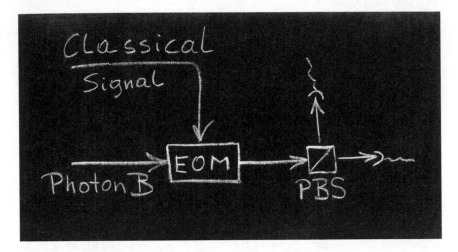

Figure 45. Bob's experimental setup in the Danube teleportation experiment. Out of the glass fiber arrives his photon B, which was initially entangled with Alice's photon A. The classical channel, which is a microwave radio link, tells him which entangled state was measured by Alice. In one of the two cases, Bob's photon B is immediately in the original state of the original photon X, and Bob has nothing he needs to do. Teleportation has succeeded. In the other case, the polarization of Bob's photon must be rotated. This is done using the electro-optical modulator (EOM). Its operation is basically very simple. We can apply some voltage to it. If no voltage is applied, the photon just passes through without modification. If the right voltage is applied, the photon is rotated in the desired way. We identify the state of the photon by measuring its polarization using the polarizing beam splitter (PBS). The PBS can be rotated around the beam axis in order to identify any linear polarization. The confirmation of the fact that the teleportation works is that only the correct one of the two detectors behind the beam splitter registers the photon and never the other. It must be the detector that corresponds to the initial polarization of the teleportee photon X.

"Finally, we measure the polarization of the outcoming photon. This confirms that the teleportation procedure did work, and then the experiment is finished."

Rupert pauses and, apparently not convinced we have understood everything, continues. "What happens in our teleportation experiment is actually quite simple. We prepare different polarizations at Alice's input, and then simply prove that the photon coming out at Bob's side always has the same polarization that we set up at Alice's input. Note"— and he points at the drawing of the setup (Figure 42)—"that it is not the original photon X which arrives at the output. It's actually Bob's twin member of the entangled pair, photon B. The important point is that Alice's original photon X completely loses all its polarization features when it becomes entangled at the fiber coupler. After all, a fully entangled particle has no private features. In our case, it does not have any polarization. The polarization feature is teleported over to Bob's photon in two ways, using the quantum channel and the classical channel. The original is destroyed, but we end up with photon B on the other side of the Danube being exactly in the state the original had. So we have achieved teleportation.

"Like every experiment, our experiment has its shortcomings, because the real world is not as perfect as the ideal imagined by our theory colleagues," Rupert adds. "First, we detect only about 30 percent of the photons, because our detectors are not perfect. No detectors are perfect. But that's something we can live with. It simply means that our teleportation device is actually better than we are able to measure. A second problem is that we are able to identify only two of the four entangled states. So, our device only operates in 50 percent of the cases, because which entangled state appears is completely random and we cannot influence that. That means that only half of the initial photons are really teleported, and half get lost. But anyway, and most important, when we identify one of the two entangled states, our teleportation procedure has succeeded. The quality of our teleportation procedure is not reduced because we miss some photons or because we are not able to identify all entangled states. What is reduced is the fraction of cases where teleportation succeeds. It's up to future graduate students to improve the experiment to the satisfaction of every critic. But that," he says, smiling, "will probably never happen, because people always like their own experiments better than those of others. Well, not always."

THE MULTIPHOTON SURPRISE AND, ALONG THE ROAD, QUANTUM TELEPORTATION

It took more than four years after the 1993 proposal by Charles Bennett and his colleagues for the first teleportation experiments to occur. This was done in 1997 in Innsbruck. You might say that this is quite a long time, but given the technical challenge, it is rather short. When the teleportation proposal appeared, we were sure that it would take many years to realize it. Yet, without knowing it, we were already developing the necessary tools for a completely different reason.

Actually, for the experiment, a number of completely new challenges had to be tackled. One was, What would be a suitable source for the entangled particles? Would photons be the best? There had been some entanglement experiments mostly with photons, but really good sources did not yet exist at the time of the theoretical proposal. Even more important, no one had any idea at the time how to perform a Bell-state measurement, or how to identify an entangled state between two independent photons by measurement. And most important, no one had done an experiment on more than two photons or on any two entangled particles before. There had been a number of nice experiments on Bell's inequality, but they had all been two-photon experiments. So in view of these challenges, the time it took until the first experiment was actually short. As often, serendipity was crucial. In our case it was the fact that my group and I had been interested in similar experiments for quite some time.

Luckily, my group was already very interested in performing experiments with three (or more) photons before the idea of teleportation came about. In 1986 Dan Greenberger of the City College of New York visited me for a couple of months when I was still in Vienna. When he came, we thought about what a good research project to collaborate on during his stay would be. It turned out that both of us had been asking ourselves how one could go beyond conventional entanglement experiments. We realized that all the experiments and the theory thus far had been limited to entanglement between two particles only. So, why not consider entanglement between more particles? We actually started to work on the entanglement of four particles because we could not conceive of a way to create the entanglement of three. The idea back then was to have an initial particle that decays into two, with each of the two then decaying into two.

When we looked at the theoretical predictions for the correlations of these four photons, we got a huge surprise. The mathematical expressions of correlations between four particles are very complicated because there are so many variation possibilities. Each of the four can be subject to many different measurements. Instead of measurements on two particles, the experimentalist carries out measurements on four particles, and the measurements depend on the settings of all four.

The mathematical results for measurements on a four-particle entangled state, though correct, were a complicated mess. Therefore, Danny Greenberger and I decided to focus on a small subset of the theoretical predictions first, those for perfect correlations. Remember, perfect correlations like those between identical twins were the starting point for John Bell when he derived the contradiction between local realism and quantum physics. The conflict of quantum physics with that philosophical view then appears only for the statistical predictions of quantum mechanics, while local realism is very well able to explain the perfect correlations. From a philosophical view, this was a rather comforting situation. After all, perfect correlations are the domain of classical physics. That is, there we can make predictions with certainty once we know enough features of a system. And in principle, it is not so unexpected that such a classical viewpoint would break down for the statistical predictions of quantum mechanics, since after all quantum mechanics is a statistical theory. But in the four-particle entanglement case, which we had started to study, the perfect correlations had a big surprise in store for us!

Surprisingly, these perfect correlations are self-contradictory!

We had already found at that time that three-particle entanglement would run into the same surprising situation, but we did not focus on three particles yet, as we had no idea how to produce three-photon entangled states. This is something we learned later. So, let us assume we have a source emitting three-photon entangled states (Figure 46). This is called a GHZ source, since in the scientific community, the kind of entanglement we are talking about here has been named GHZ entanglement after the initials of Danny Greenberger; Mike Horne, with whom we had a long-distance collaboration, as he was still back in the United States; and myself. The most basic such state is a superposition of all three photons horizontally polarized (HHH) or vertically polarized (VVV) along some direction. Before measurement, none of those photons carries any polarization. If any of them is measured with a two-channel polarizer oriented along that direction, it randomly assumes H or V polarization. Then the other two photons immediately obtain the same polarization state. This is Einstein's spooky action taken one step further.

We can easily imagine many other possible polarization measurements that could be performed on these three photons. We could use polarizing beam splitters, or two-channel polarizers that can be rotated around the respective incoming beams. So, we could measure vertical and horizontal polarization along any possible direction. We also could decide to measure circular polarization.

For most combinations of the three polarization measurements, the correlations are just statistical. But for quite a number of measurements, perfect correlations occur. This means the following: measurement results on each of the photons are completely random. So, for example, H or V appears with the same fifty-fifty chance, independent of the orientation of the respective polarizer. But, if we know the polarization measurements of two photons, the polarization of the third photon can be predicted with certainty if the orientations of the three polarizers fulfill certain conditions. One such condition is that the orientation angles of two polarizers have to add up to that of the third. This is more general than the case of two photons, in that the three polarizers need not be oriented in parallel anymore.

The interesting point now is the following. There are sets of polarizer orientations that demonstrate a complete contradiction between local realism and quantum mechanics in the following sense. Suppose we know the polarization of the first photon to be H' (rotated by 45 degrees from H) and the polarization of the second photon also to be H'. Then, a local realist would predict the third photon to be H'-polarized also,

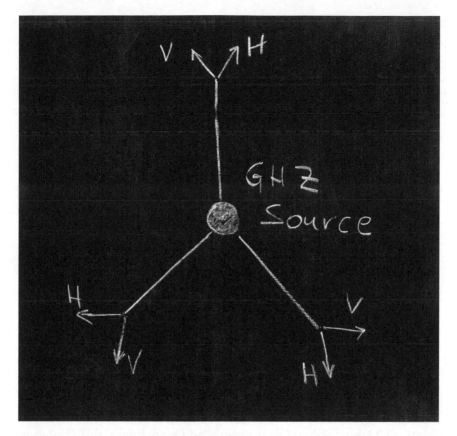

Figure 46. Essence of a GHZ experiment. Consider an entangled state of three photons. Each one's polarization can be measured. The essential point is the following. When the orientations of the three polarizers fulfill certain conditions, then on the basis of the random results for two photons, the polarization of the third photon can be predicted with certainty. For example, if the first two photons are both H-polarized, quantum mechanics would predict the third photon to be V-polarized. Interestingly, a local realist would make the opposite predictions. He would predict that the third photon must be H-polarized. Experiment confirms the predictions of quantum mechanics. Thus, as opposed to the situation with two-particle correlations, local realism is not even tenable when quantum physics makes predictions with certainty. This is the maximal possible conflict between the two views of the world, the quantum one and the local realist one.

while quantum mechanics would predict it to be V'-polarized. This is the maximal possible contradiction between the two worldviews. Both quantum mechanics and local realism make definitive predictions for that situation, but these predictions are completely opposite each other. Quantum mechanics is at variance with any local realistic picture, even for perfect correlations, even for each individual photon! Or in another sense, the argument of Bell cannot even get off the ground, because the perfect correlations in two-particle systems were the starting point of his reasoning. Local realism is not even able to correctly describe those situations where quantum physics allows us to predict a measurement result with certainty!

Ever since Greenberger, Horne, and I had discovered this contradiction in 1987, it had been my scientific goal to verify these correlations experimentally. My group and I had a formidable task in front of us. We had to develop new experimental tools for entanglement between more than two particles—how to create such entanglement, how to measure it, how to handle these states in experiments, and so on.

This was completely uncharted territory. Not only did we have to develop new experimental tools and components, we also had to develop a new way of thinking about such kinds of experiments, as no one had ever considered entanglement beyond two particles before. So it took us eleven years until, in 1998, Dirk Bouwmeester, Jian-Wei Pan, Matthew Daniell, Harald Weinfurter, and I were finally able to observe three-photon entanglement in the laboratory in Innsbruck and to confirm the quantum mechanical predictions.

In order to achieve that goal, we developed a number of instruments that turned out to also be important in quantum teleportation. Important outside help came from my colleague and friend Marek Zukowski of the University of Gdansk in Poland. We had numerous discussions with him, during which we developed ideas for the experimental realization of entangled states with three or more photons. Actually, we had to abandon many ideas because most of them did not work. But in the end, we found a solution.

One problem is that it was not possible to produce entangled states of more than two photons in a direct way. You have to start from two-photon entanglement. But how do you create higher-order entanglement out of pairs? Our idea is in principle very simple. You first create two entangled pairs, that is, four photons. Then you measure one photon in such a way that you do not know, even in principle, which of the two pairs this specific photon came from. Then the remaining three photons are entangled. For the experiment, we also had to develop many other technologies, such as ways to effect the precise timing of the photons, to create two pairs using the process of spontaneous parametric down-conversion, and to identify entangled states using beam splitters, as well as polarizers and so on. These were only a few of the challenges.

While working toward preparing that experiment, we soon realized that the tools we developed could also be used to do quantum teleportation. The teleportation experiment was finally done by Dirk Bouwmeester, Jian-Wei Pan, Klaus Mattle, Manfred Eibl, Harald Weinfurter, and myself. In that first experiment, and in all others since then, the signature confirmation of successful teleportation has been to prove that whatever state is prepared by Alice ends up on Bob's side. This does not have to be shown for every conceivable quantum state because there are so many. But on the other hand, it is not sufficient to show it just for horizontal and vertical polarization, because the possibility that the apparatus has some preference for one basis cannot be ruled out. So, in the experiments one also has to demonstrate teleportation of superpositions of the horizontal and vertical, such as linear polarization rotated by 45 degrees or some circular polarization. This was achieved in 1997.

In that experiment, the distance of teleportation was only of the order of about one meter. But this distance has significantly increased by now. For example, in the Danube experiment, the distance is 600 meters (2,000 feet). After that first successful experiment, we turned to the teleportation of entanglement and to the realization of three-photon entangled states already mentioned.

TELEPORTING ENTANGLEMENT

Thus far, we have learned that we can teleport the state of a photon or another particle. This means transferring properties of the original particle over to another particle. But what happens if the particle to be teleported itself is entangled? Remember, we learned that entanglement means that the particle does not have its own state; it does not carry any properties of its own.

Let us analyze the experiment in detail (Figure 47). We start with two entangled pairs. A and B are entangled with each other, and X and Y are entangled with each other. Then, we take one photon from each pair—A from pair A-B and X from pair X-Y—and submit them to a Bell-state analysis (BSA) procedure, which entangles those two coming from different pairs.

Remember that a particle's state is what can be said about the particle (more precisely, what can be said about possible future measurement results on the particle). So, let us assume that all that can be said about particle X is teleported over to particle B. But what can be said about particle X? It has no state of its own. All that can be said about it is that it is entangled with Y. Therefore, we conclude that the same can now be said after the Bell-state measurement about particle B. Particle B and particle Y end up entangled even though they share no common past!

In the original teleportation experiment, there was also a classical channel. This channel is not shown in Figure 47, but it is also necessary. Why is it necessary here? As in the original teleportation experiment, the Bell-state measurement on photons A and X can have four different results, representing the four possible entangled states. In that experi-

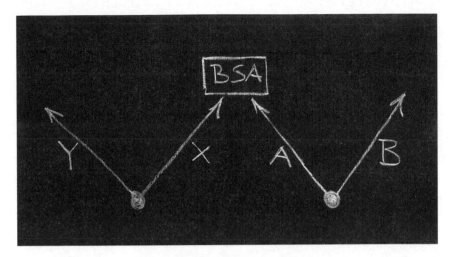

Figure 47. Teleportation of entanglement. Initially, we have two entangled pairs, A-B and X-Y. Then, just as in regular teleportation, we perform a Bell-state analysis (BSA) on two of these photons, A and X, one from each pair. What then happens is that the entanglement X shares with Y is teleported over to B. Alternatively, we could also say that the entanglement A shares with B is teleported over to Y. Regardless of the way we look at the situation, the result is the same. The two photons B and Y end up being entangled. This is rather remarkable, since B and Y had no connection whatsoever in the past. They were created completely independently. So, two systems that had never interacted with each other or shared a common past can be entangled.

ment, this meant that Bob's photon B could be in four different quantum states, all uniquely related to the state of the original. It now means that Y and B can be entangled in four different ways. Y and B can be in any of the four possible entangled states. The specific entangled state in which photons B and Y end up is just the same as Alice obtained randomly for A and X. For any use of that entanglement, we must know the nature of the entanglement. Therefore, Alice's measurement result has to be communicated to whoever wants to utilize the new entanglement between B and Y. This could, for example, be Bob receiving particle B, or it could be someone else receiving particle Y. So, in principle we need at least one classical channel in the experiment.

That experiment was proposed in 1993 in a joint article by Marek Zukowski, Mike Horne, Artur Ekert, and myself. At that time, we called that protocol *entanglement swapping* because the initial entanglement between A-B and X-Y was swapped over to A-X (Alice's measurement result) and B-Y.

Conceptually, the most interesting part here is the entanglement of the two outer photons—Y and B—which share no common past. They neither come from the same source, nor have they ever met. It turns out that the conventional picture is wrong, namely, that entanglement is something created when two systems interact with each other or when they are produced together in some way. Entanglement might be created in one of these ways, but this is not necessary. Many physicists hold the view that entanglement is a consequence of some conservation law, for example, the conservation of angular momentum or the conservation of energy. As we can see now, obviously, this is not a necessary precondition for observing entanglement.

The experimental realization of the teleportation of entanglement, or entanglement swapping, was done by Pan, Bouwmeester, Weinfurter, and myself again in Innsbruck in 1998 shortly after the first experiment on teleportation. The experiment clearly confirmed that entanglement itself can be teleported.

In these experiments, the quality of teleportation was not perfect. Sometimes, the polarization correlations between B and Y were the wrong ones. But there is a definitive proof—the prediction that photon Y and photon B must be entangled so strongly that Bell's inequality will be violated. That is the definitive proof of entanglement. But in our first experiments in 1998, the quality of the teleported state was not good enough to actually see such a violation of Bell's inequality.

A few years later, in 2001, Jennewein, Weihs, Pan, and myself had improved our experiment so much that the quality of the teleported photons was much, much higher. So we set out to prove exactly that. We were indeed able to demonstrate that photon Y and photon B are so strongly entangled with each other that they violate the Bell inequality. This means that two photons that never interacted with each other in any way are now connected by Einstein's spooky action at a distance. This experiment also definitively proves the quantum nature of teleportation.

Let us recall for a moment the experiments we just considered. Upon Alice's measurement of her two photons X and A, Bob's photons, which are measured at a later time but never interacted with each other, become entangled. But Alice could also decide not to perform such a measurement. Then, Bob's photons Y and B would not be entangled with each other. So it is her decision whether or not to perform a measurement that decides whether the other two photons become entangled or not. This might sound strange, but a much funnier situation is possible.

A GHOSTLY IDEA

In the year 2000, the late Asher Peres from the Technion in Israel, a pioneer of quantum teleportation theory, had a rather strange, surprising, and elegant idea. He suggested the following: First, Bob measures the polarizations of photons B and Y (Figure 48). Then, once the measurement results on these photons already exist and are registered, say, in the computer memory or even written down on a piece of paper, Alice performs her Bell-state measurement on her two photons X and A. Upon Alice's measurement, Bob's photons B and Y become entangled. Even stranger, Alice could decide not to perform a Bell-state measurement on her photons A and X, and then Bob's photons B and Y would remain unentangled. So, Alice can decide at a point in time when Bob's photons Y and B no longer exist, when their polarizations have already long ago been measured and the results written down somewhere, whether these photons are entangled or not.

How can that be? How is this possible? Certainly, Alice's measurement cannot act back into the past and influence the earlier measurement results on Bob's photons Y and B. The measurement results

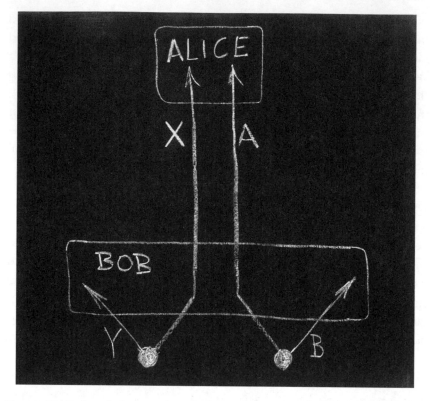

Figure 48. Delayed-choice teleportation. Alice and Bob create two pairs of photons, A-B and X-Y. Bob measures one photon each from each pair, B and Y. The other two, A and X, are sent over to Alice, who may decide at a later time, long after Bob has measured his photons, which measurement she will per-form on the two. One of the choices she has is to entangle the two with each other, just the way it is done in quantum teleportation. That way, the outer two photons, B and Y, become entangled with each other long after they have been measured! Another possible choice is that Alice might decide to perform on A and on X separate measurements of their polarizations. That way, B and Y do not become entangled with each other. Rather, the data indicate the en-tanglement between A and B and between X and Y. In that case, the results obtained by Bob much earlier for B and Y have a completely different mean-ing. What this experiment tells us is that there is an essential difference in quantum physics between the observed individual results and the interpreta-tion that we assign to these results. The measurement results on B and Y exist prior to and independent of Alice's decision. But without Alice's measurement results, they do not have any meaning; they are completely random. Their meaning, their interpretation, is assigned to them by the kind of measurement Alice decides to perform on A and X. Alice's choice of what to measure can be delayed by any amount of time.

have already been registered, maybe even written down on a piece of paper. Changes of written-down measurement records certainly do not happen. But what happens is no less interesting. Actually, philosophically, this circumstance gives us a very important message, as we will see now.

Let us carefully analyze what we expect. At first, we look at Bob's results. Well, we realize that both his photon B and his photon Y are initially entangled with other photons. B is entangled with A, and Y is entangled with X. We remember that entangled photons do not have a polarization before they are measured, and we also remember that upon measurement, they randomly assume some polarization. So we conclude that Bob's measurements of his photons Y and B each result in sequences of random results or random sequences of 0 and 1. Now we may sit down and scratch our heads for a moment. What do these random-number sequences mean? How are they to be interpreted?

We will now see that the individual results of Bob's photons Y and B can acquire very different meanings, depending on which measurement Alice decides to perform. She may decide to measure the polarization of photon X and the polarization of photon A, each by itself. What do we expect? We know that X is part of the entangled pair X-Y and A is part of the entangled pair A-B. Like Bob's results, Alice's results will also be sequences of random numbers, a sequence for photon X and a sequence for photon A. These two sequences of random numbers will be completely independent of each other, since A and X were created independently from each other. But the results for each photon will be strongly correlated to the results obtained for its respective twin photon. So the correlation between Alice's result for A and Bob's result for B will confirm that these two photons were produced in an entangled state. Likewise for the pair X-Y.

The data will, for example, result in a violation of Bell's inequality both for the A-B pair and the X-Y pair. Whatever Alice does, she will definitely conclude that A and B are perfectly entangled and that X and Y are perfectly entangled. Alice and Bob together will also conclude from their data that B and Y have nothing to do with each other. They are completely independent and unrelated.

On the other hand, Alice could decide to perform on her photons A and X a joint Bell-state measurement. We remember that this is just the measurement done in quantum teleportation. It entangles A and X.

This means that Bob's photons B and Y now become entangled also. But wait a minute, these two photons have already been registered by Bob, and the measurement results have already been written down on a piece of paper or stored in a computer. How can that be? How can the measurement results now reflect that B and Y are entangled, just because Alice decided to perform a Bell-state measurement on A and X? Even though before, when Alice measured her photons separately, Bob's photons B and Y were not entangled, their measurement results were completely uncorrelated? How is that possible?

The solution to our puzzle actually is very exciting. Let us first look at the case where Alice performs a joint Bell-state measurement on her photons X and A. Then, she meets Bob, and they try to make sense out of his data, which he obtained by measuring his photons Y and B. We now remember that there are four possible results for Alice's Bell-state measurement. This means that in her measurement, not just one entangled state occurs all the time, but rather, four different entangled states. Which specific state occurs in an individual measurement of a specific pair is completely random. There is no rule about which state will occur when.

So, what Alice and Bob do when they meet is, they sort Bob's data obtained earlier on Y and B into four subsets, into four different bins — one bin for each of the four entangled states obtained by Alice. And then it happens that each of the four sets obtained by Bob on B and Y confirms that these two photons have been entangled with each other, even when they were measured earlier. In each of the four bins, Bob's photons are entangled in a different way, namely they are in exactly the same entangled state that was obtained randomly by Alice for her two photons. While each of the bins indicates a specific entanglement, if we mix together all four bins, the results all together are completely random and do not indicate any entanglement. So Alice's results allow us to sort Bob's data into the right sets, which individually are not completely random anymore, while the complete set was.

On the other hand, let us look at the results Alice obtains when she measures the polarizations of photon X and photon A individually. That way, these two photons are not projected onto any Bell state. Rather, X remains entangled with Y, and A remains entangled with B. Now, Alice and Bob, when they get together, will sort Bob's Y data according to which results Alice obtained for X and the polarization she measured, and likewise, they will sort Bob's data for photon B according to the polarization Alice measured on A and the result she obtained. Now, these data sets will perfectly confirm that Bob's photon B was completely entangled with Alice's photon A, and Bob's photon Y was completely entangled with Alice's photon X. The important point is that this sorting of Bob's initially random data results in completely different subsets for his data as compared with before, when Alice did the Bell-state measurement on her two photons. Thus, the same data that earlier confirmed entanglement between Y and B now confirm that Y is entangled with X and B is entangled with A.

Philosophically speaking, we have a very interesting situation. The data obtained by Bob long before Alice decided what kind of measurement to perform can be part of two completely different physical stories. The specific physical picture depends on Alice's later measurement. In a sense, the data have no story to tell before Alice makes her decision and does her measurement accordingly and this decides the meaning of Bob's data. One might very well say that Bob's data are a primary reality in no need of explanation. If we wish to have an explanation, we need to complete the experiment. This completion of the experiment requires Alice to make a decision that defines the meaning of the data already obtained.

The message to be learned is that individual events in quantum physics are primary: they are more fundamental than the explanations that we later construct based on our physical pictures. Bob's individual events just happen, independently of whether his photons are entangled with Alice's photons or not. We also note that a crucial ingredient in our discussion is the objective randomness of the individual results. It is the fact that neither Bob nor Alice can influence which specific result they obtain for a given photon. This randomness prohibits any signaling back into the past, as might be possible if Alice could influence her outcome. It also guarantees that very different physical explanations for the same random result can be obtained by us.

The idea of delaying Alice's measurement to a time until after Bob has measured his two photons was realized experimentally by us in Vienna in 2001. The experiment was again performed in optical glass fibers. The Bell measurement was done using a fiber coupler, just as in the experiment at the river Danube. In order to delay the Bell measurement to a time after the two outer photons had already been registered by Bob, Alice's two photons each went through a ten-meter-long glass fiber before entering the fiber coupler. The experimental results clearly confirm that the photons Y and B registered earlier by Bob can be seen as entangled, if sorted the right way, on the basis of Alice's measurement results.

In that experiment, in principle, Bob's earlier measurement of his two photons could have influenced through some unknown communication the measurement results at Alice's station. In order to exclude such a hypothetical possibility just as in the Bell experiments, Xiaosong Ma in my group, together with Stefan Zotter and Thomas Jennewein, in 2009 performed an experiment where Alice's choice of whether to project her two photons onto an entangled state or to measure them separately was made by a quantum random-number generator (QRNG), space-like separated from Bob's measurements. This experiment again fully confirmed the quantum predictions, and it rules out in this experiment, too, any possible explanation by an unknown communication.

CONNECTING QUANTUM COMPUTERS

Such experiments are not only of philosophical interest. Many people feel that they will have important technical applications in the future. One basic idea is to connect future quantum computers by using entangled states. In general, the output of the quantum computer is some kind of quantum state. Let us assume that this output is now required as an input to another quantum computer at some distance. It would be ideal to teleport the output state of the first quantum computer over as input to the second. If the quantum computers are widely separated, maybe in two cities, we would need to be able to teleport over large distances.

One problem with teleportation over large distances is the simple fact that photons may get lost along the way. With glass fibers, we can cover about one hundred kilometers (sixty-two miles) at most. A similar limit for distance holds for the transfer of photons through air. So, how can we span larger distances? One possibility is the kind of experiment we just talked about, namely, the teleportation of entanglement, or as we called it earlier, entanglement swapping. Through chains of such teleportations (Figure 49), we could cover much larger distances.

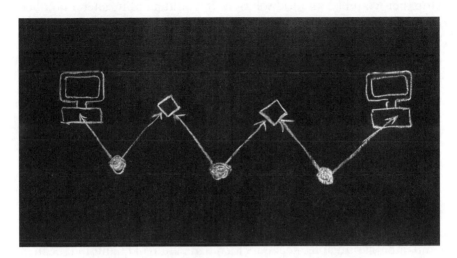

Figure 49. By means of quantum repeaters, two widely separated quantum computers may be connected to each other. In principle, this is multiple tele-portation of entanglement. In addition, there could be at the intermediate stations small quantum computers that purify the quantum states.

It would be ideal at the intermediate stations to not just perform Bell-state measurements but to also correct for photon loss. A quantum amplifier would be an ideal solution. It would amplify, after a number of kilometers, the incoming quantum signal to cover larger distances, just as amplifiers are used today in all long-distance communication cables, including glass-fiber connections. But the problem is that quantum states cannot be amplified. This is, as we have already noted, because quantum states cannot be cloned. Therefore, the intermediate stations used in a chain of teleportation, as shown in Figure 49, would not be amplifiers of a quantum signal, but something like repeaters, which would achieve two objectives. First, they would perform a Bell-state measurement and therefore teleport the entanglement over a larger distance. Second, very small, mini quantum computers could purify the incoming quantum states, correcting for transmission errors and, using entangled states in parallel, make up for losses. Some of the fundamental principles necessary for a realization of quantum repeaters have already been demonstrated in experiments. However, a complete repeater station is technically not feasible yet.

REALITY VS. INFORMATION

The teleportation of an entangled state is probably the clearest proof of the quantum nature of teleportation. It furthermore throws a bright light on the relationship of our description of nature with physical reality. Let us recall again the situation shown in Figure 48. It is a fact that Bob's registration events, that is, the results he obtained when measuring his photons Y and B, are objective ones. That is, they are written down in some way; they exist; everyone can look at them, and everyone can agree on what these results are. Furthermore, they are not in need of an interpretation. They are just events and that's it. But we physicists want to understand these events. We want to describe why these events happen. So we must present an explanation, a consistent interpretation, and that is where an interesting question now arises.

The interesting point is that in the end we will, for Bob's results, present a different interpretation depending on what Alice at a later time decides to do. She may decide to do a Bell-state measurement, or she might decide to do a measurement on each photon on its own, or there is even an infinite zoo of possibilities in between. Depending on what Alice decides to do, the results registered earlier by Bob, the events that already happened, acquire a very different meaning.

Therefore, we conclude with two important points. First, the observed events are just events, and they are in no need of any interpretation. They are, so to speak, here before we as the observers even begin to worry about what they actually mean. Second, the explanation of the events depends on later actions and decisions we or someone else might make. It is important to realize that the two interpretations of Bob's data are mutually exclusive. The reason is that entanglement is monogamous. Bob's photon Y can be entangled only with X or B but not both.

At this point, we might state one point with emphasis. Each of the interpretations we come up with for Bob's earlier results is completely correct and objective. The fact that an interpretation depends on Alice's future decision does not at all make it incorrect or subjective. Quantum physics describes all of these situations well. It is just that the mathematical description, the quantum state we assign to the situation, is very different depending on what Alice decides to do. It depends on the specifics of the experimental settings, and therefore, it depends on Alice's decisions, or in general it depends on the decisions of us as experimenters. And this may include specifics of the experiment that will be decided in the future.

These kinds of considerations support the point of view that has most succinctly been expressed in the Copenhagen Interpretation of quantum physics as it was created specifically by Niels Bohr. According to that interpretation, the quantum state of a system is not a field, or some entity that spreads "out there" in space and time. On the contrary, the quantum state is just a representation of the knowledge we have of the specific physical situation we are investigating. This representation of our knowledge naturally depends on which situation we have in front of us, which kinds of measurement results we obtain.

In our specific case, our knowledge of the specific situation depends on the kind of measurement Alice performs at a later time and the result she obtains then. We conclude that Alice's later measurement, and her result, do not influence the physical reality already in existence—namely, the specific results Bob obtained at an earlier time. But they change what we may say about the situation. They change our interpretation of what is going on. This very much calls to mind a famous statement by Bohr: "There is no quantum world. There is only an abstract quantum physical description. It is wrong to think that the task of physics is to find out how nature *is*. Physics concerns what we can say about nature."

This statement was conferred on us by Aage Petersen. There is no document by Bohr himself where this expression appears explicitly. So there is some debate about whether Bohr really said it. In my opinion, it expresses his position perfectly.

FURTHER EXPERIMENTS

Quantum teleportation is certainly not limited to teleporting the polarization states of photons, as was done in our Innsbruck and Vienna experiments. We can also teleport other properties, the states of other physical quantities, for example, the energy state or the momentum state of a photon or some quantum state of an atom. It is also possible to teleport more complex systems.

In 1994 Sandu Popescu, then at the University of Cambridge and now at the University of Bristol, proposed an idea that is mathematically equivalent to teleportation but has a different physical meaning. He suggested having the state that is being teleported carried piggyback on one of the two entangled photons. That idea was realized in 1997 by Francesco De Martini's group at the University of Rome. They used only one photon pair, which was entangled in momentum. The state to be teleported was then imprinted, by means of suitable polarizers and polarization rotators, onto Alice's photon. That kind of experiment, using two-photon entanglement only, cannot achieve teleportation of a photon state coming from outside. To do that, one would have to transfer the polarization state of an independent, separate photon onto Alice's entangled one, which is equivalent to taking the Bell-state measurement of two photons. Nevertheless, the Rome experiment is rather interesting. A more suitable word probably might be "telepreparation" instead of "teleportation." This is because in the end, the preparation performed on Alice's photon is transferred over to Bob's. Harald Weinfurter's group at the University of Munich recently performed some other experiments on telepreparation.

In the teleportation of polarization, a discrete property is transferred.

This means that the photon has two possible polarizations, horizontal or vertical, as defined by a polarizer's orientation. But photons also have some continuous properties that are not discrete—for example, their frequency or energy. Interestingly, we can view light as a super-position of many (in theory even an infinite number of) different oscillations of the electromagnetic field. Because of the quantum nature of the electromagnetic field, its oscillation also has an associated quantum mechanical uncertainty, just like Heisenberg's uncertainty principle. This means that there is some noise in the photon's oscillation modes. Interestingly, we can reduce the noise in one mode of oscillation. The Heisenberg uncertainty principle tells us that there is always a complementary quantity whose uncertainty is then increased. Here this means that the noise in some other modes of oscillation will then increase.

States with reduced noise in some mode of oscillation are called "squeezed" states. Some researchers hope that more precise mea-surement of various physical quantities in the future might be possible using such squeezed states. One idea, for example, is the eventual ob-servation of gravitational waves as predicted by Einstein's general theory of relativity. In the experiments, the observation is performed by mea-suring tiny motions of mirrors suspended in space rather accurately, and the hope is to be able to observe the position of such mirrors using squeezed light better than what is possible today.

Squeezed light has also found its application in teleportation. The group around Jeff Kimble at the California Institute of Technology suc-ceeded in 1998 in teleporting the squeezing property from one beam of light to another. The experiment follows in essence the initial propo-sal for teleportation, which has to be translated into the language of squeezed states of light.

In 2001, the group around Eugene Polzik, then at Aarhus Univer-sity, Denmark, and now in Copenhagen, succeeded in entangling the joint squeezed state of the spin of many cesium atoms within an atomic cloud with the state of the spin in another atomic cloud. The atoms form an ultracold cloud that is trapped in some arrangement of electri-cal and magnetic fields. The quantum mechanical property of spin is a generalization of angular momentum. While an ice-skater can turn only either to the right or to the left (Figure 50), an atom can actually exist in a superposition of both possibilities at the same time. The spins of many such atoms can be aligned with respect to each other to a certain degree. The stronger the alignment of the spins, the more squeezed the state of spin of the atoms will be. In 2004 Polzik's group succeeded in actually teleporting the squeezing property from one cloud of atoms to another. In 2007 the same group was able to achieve teleportation between photons and atoms.

Another interesting step forward occurred in 2004, when two groups—one around Rainer Blatt at the University of Innsbruck and the other one around David Wineland at the National Institute of Standards and Technology (NIST) in Boulder, Colorado—independently reported the teleportation of states of atoms. Specifically, this does not mean the teleportation of the whole quantum state of an atom, but the teleportation of one particular substate of an individual atom. In both experiments, people used electrically charged atoms, called ions, that were trapped in a suitable arrangement of electromagnetic fields.

In principle, it is possible to keep individual ions in these kinds of traps for an indefinite time. They are well protected from the environment, and therefore, the quantum states have a long life. Using lasers, we could write information onto the ions and read it out again. The group in Innsbruck used calcium atoms, and the NIST group beryllium atoms. The entanglement between the quantum states of two atoms is the quantum channel. The third atom is then the one whose state is teleported. By means of a Bell-state measurement on that atom and on one of the two entangled ones, its quantum state is transferred over to the other entangled atom. The distances over which teleportation is realized in these experiments are small, well below one millimeter, because the atoms are sitting in atomic traps close to each other.

Seen in a broader perspective, all teleportation experiments are part of a research program to realize quantum computers using quantum systems.

A most interesting feature of a quantum computer is the representation of information in qubits. For example, the information carrier in a quantum computer could be the spin of an atom: we could say that spin corresponding to clockwise rotation is one kind of information, say, the 0 of a qubit, and rotation in the opposite direction is another kind of information, say, the 1 of a qubit. Then, if the two spins were entangled, the information carried by the two spins would be entan-

Figure 50. Quantum pirouette. According to the laws of classical physics and certainly also in our daily life, an ice-skater can spin only either to the right or to the left. Yet, a quantum ice-skater could exist in a superposition of both possibilities, as shown. For ice-skaters, this is just fantasy, but for atoms and other quantum particles, this is reality in experiment.

gled. So, information in a quantum computer would not only be entangled, but it would also exist in superpositions of many possibilities. Interestingly, there are a number of problems a quantum computer would be able to solve that any conventional computer would take much longer than the age of the universe to solve. What does that have to do with teleportation? We mentioned that teleportation is the perfect way to transfer quantum information from the output of one quantum computer to the input of another. Quantum teleportation can also be used to process quantum information within a quantum computer itself. To create a quantum computer, we need in principle the same kinds of methods that we need to realize quantum teleportation. Indeed, Michael A. Nielsen and Isaac L. Chuang in 1999 proposed a quantum computation scheme where teleportation plays a key role inside the computer.

Therefore, since quantum teleportation works so nicely for photons, it is reasonable to expect that it should also be possible to build a quantum computer working according to the same principles. Such a quantum computer would use only photons, not ions or atoms. Then the carrier of the information would not be anything material, since photons do not have any rest mass. That this is indeed possible was shown in 2001 by Emanuel Knill and Raymond Laflamme at Los Alamos National Laboratory and Gerard Milburn at the University of Queensland. Since then, a number of elementary elements of such all-optical quantum computers based on photons have been realized in experiments.

Such quantum computers operate according to the principle of chance. This means that, at the end, the computer only occasionally shows the result of the calculation. In many cases, it ends up in a state that is not useful for the computation intended. Also in 2001, Robert Raussendorf and Hans J. Briegel, then of the University of Munich, showed that it is in principle possible to build a quantum computer where the problem of chance is circumvented.

The first demonstration of such a *one-way quantum computer* was performed in an international collaboration at the University of Vienna in 2005. My collaborators in this work were Philip Walther, Kevin J. Resch, Markus Aspelmeyer, and Emmanuel Schenck on the experimental side in Vienna and Vlatko Vedral and Terry Rudolph, our theory colleagues, from Imperial College in London. One big advantage of such a one-way quantum computer is that it is much faster than any other quantum computers considered today. Therefore, in principle, it should be possible to build future quantum computers that are based on photons, that is, quantum light only. This is a specifically useful approach for small quantum computers performing simple operations.

QUANTUM INFORMATION TECHNOLOGY

In this book, a hidden story has been developing in front of our eyes. It is a story that has happened many times in the history of physics specifically and of science in general. It always starts with the fundamental motive for doing science, namely, the curiosity of an individual. In the case of entanglement, these individuals were Einstein and Schrödinger in the 1930s, as well as other founding fathers of quantum mechanics. They wondered about some of the predictions of quantum mechanics that, when applied to individual systems, to individual particles, lead to counterintuitive phenomena. One of those predictions was entanglement. Another one was the curious finding that we have objective randomness in quantum physics, the fact that randomness plays a fundamental role in the theory and is not just a measure of our ignorance. While these features, including quantum superposition, were known mathematically to many who were working in early quantum mechanics, it took great minds like Einstein's and Schrödinger's to realize how peculiar the features are in terms of their philosophical applications and their meaning for our picture of the world. Remember that Einstein did not want to accept randomness, saying that "the old fellow does not play dice," and Schrödinger suggested that entanglement is *the* essential feature of quantum mechanics.

The next step in our story was then an interesting coincidence. In the mid-sixties, John Bell discovered that an important philosophical question raised by entanglement can actually be tested in experiment. This was the question of whether or not nature can be described in what is called a local realistic picture, a picture with no place for what Einstein called "spooky action at a distance." The coincidence was that around that time, the laser was invented, which made experiments testing local realism possible. This was not the situation in the 1930s, at the time when Einstein and Schrödinger raised their philosophical questions. So the experiments that began in the 1970s and that confirmed the predictions of quantum mechanics rather than those of local realism were actually motivated by philosophical questions, or in other words by the curiosity of a handful of people. This curiosity is an important motive for human endeavor, which in the sciences has often resulted in interesting discoveries when combined with new technology.

The third piece in our story was not expected by anyone who was part of the early experiments. In the 1990s these fundamental quantum ideas gave rise to ideas about new ways to transmit and process information. These new concepts include quantum cryptography, quantum random-number generators, quantum teleportation, and the quantum computer. It is absolutely certain that these ideas would not have emerged without the earlier research, which was motivated just by philosophical curiosity.

This latest part of the story is unfolding right now. The development of a new quantum information technology is one of the hottest areas of research worldwide. Many groups in many countries are working to develop quantum cryptography, quantum computers, quantum communication, and many other ideas leading to possible technological applications.

In terms of technical maturity, the most mature concept is that of the quantum random-number generator. Such devices make use of the randomness of the individual event in quantum mechanics, a feature Einstein did not like at all. Quantum random-number generators produce the best possible sequences of random numbers, which have broad applications in many diverse areas. An obvious application is in lotteries and games of chance. Suppose you bet on some roulette game on the Internet. Everyone implicitly relies on the fairness of the machine behind the scenes to produce the numbers on which people are betting. It is obvious that the quantum random-number generator would be the best way to produce these numbers, since we would then actually know that there is no hidden reason why in a certain turn of the electronic "wheel," for example, the number 21 shows up and not the number 17.

Possible applications of quantum random-number generators go far beyond games of luck, though. An important application is the encoding of secret information stored on a computer. Suppose, for example, that for reasons of national security, we want to store people's personal data on a computer for a long time. But let us also suppose that in order to protect the rights of individuals, we want to make sure that nobody has access to these data, not even in principle, unless properly authorized, for example by an independent judge. Then, the most secure procedure would be to use quantum random numbers to encode the data in computer storage and make sure that access to the random numbers is only possible if authorized by the judge. It would not be possible for anyone to read the confidential data without that authorization. There are other possible applications of random numbers, such as optimization algorithms, but we need not go into detail here.

As we mentioned, another important application of quantum ideas is quantum cryptography. People can encode secret messages and send them over to the recipient using quantum methods. The status of quantum cryptography is also rather far advanced, and many people are working on developing this new technology. For example, a large European collaboration supported by the European Commission in 2008 demonstrated a real quantum network with many different nodes across the city of Vienna, something akin to a quantum Internet.

The development of the quantum computer probably will take somewhat more time. There are some skeptical voices, expressing the opinion that the problems in the construction of the quantum computer are much too large to expect a working machine soon. I believe that we should be more optimistic here and not underestimate the creativity of experimental physicists.

We might actually be very optimistic and hope that someday *all* computers will be quantum computers. When we look at the status of information technology today, we see computer chips becoming faster and faster and able to store more and more data. This development is reflected in what computer technologists call *Moore's law*, a law formulated by Gordon Moore, one of the founders of Intel Corporation. Moore's law says that the number of transistors in a computer chip doubles every one and a half to two years. This simply is a consequence of the fact that the individual elements inside a chip, such as the individual transistors or other electronic components, get smaller and smaller as a result of technological progress.

What does Moore's law have to do with the quantum computer?

Well, Moore's law means that fewer and fewer atoms or electrons are necessary for the physical realization of an individual bit. Most interestingly, if we project Moore's law into the future, we see that in about twenty or maybe thirty years, the fundamental limit for conventional computer chips will be reached. At that point, a single bit will be represented by a single electron. This means that just as a result of the development of computer chips, the quantum limit will be reached. It could well be that technological development will slow down when chips get close to the quantum limit. But this would only mean some delay. It is therefore very reasonable to expect that eventually even conventional computers will by themselves enter the realm of the quantum.

THE FUTURE OF QUANTUM TELEPORTATION

What is the future of quantum teleportation? It is clear that within the next few years, teleportation experiments will extend over larger distances than they do today. One of the ideas various groups, including our own in Vienna, are working on is to teleport the quantum state of photons from a ground station on Earth up to a satellite, or from a satellite down to Earth. This should in principle be possible, since the atmosphere is only a few kilometers thick. A photon traveling from a ground station up to a satellite moves essentially through empty space, through a vacuum, after it has left the atmosphere behind. Traveling through empty space does not pose any problem for the photon.

Nevertheless, such experiments will be rather challenging because it will not be easy to catch a single photon that has been sent from a satellite down to Earth or vice versa. But there is no fundamental reason why such experiments should not be successful in the end. The value of these experiments will not be limited just to showing how teleportation works over larger distances. More important, they will confirm predictions of quantum mechanics on a completely new scale. In principle, two particles must remain entangled with each other, no matter how far apart they are separated.

It is interesting to contemplate whether such entanglement also might hold on a cosmic scale. Some people expect that many photons, or maybe even more complicated systems, have been entangled from the beginning of the universe. It is certainly a fascinating thought that measurement of a photon here on Earth—a photon that came to us from very far back in time—could influence the quantum state of another photon somewhere far away, even far beyond our galaxy. It is difficult to imagine how such experiments would be possible within a reasonable time, because how would we measure a photon at the other edge of our galaxy, many thousands of light-years away?

One nice idea is to imagine such experiments between Earth and a station on the Moon, or maybe even between Earth and spaceships farther away. Just consider a possible future voyage to the planet Mars. On their flight to Mars, the astronauts would essentially be bored for about 260 days. This is the time it would take to reach the orbit of Mars from the orbit of Earth. Why should these astronauts not have a little fun and play with quantum entanglement and quantum teleportation? Such experiments would extend the distances over which entanglement has been verified to tens of millions of miles.

Another line of development that will certainly give interesting results within a few years concerns the states of more complex systems, for example of atoms or molecules. The larger and more complex an object is, the larger the corresponding experimental challenges are. It is clear that the challenge rises with the number of particles making up an object. To describe a complex molecule, we have to know not only which atoms it is made of, but also how they are arranged relative to one another, how they are connected with one another. So the states we would need to teleport describing such complex molecules would have to contain a lot of information. This poses two challenges. One challenge is to create entangled states of such complicated systems, and the second one would be to invent a method of generalized Bell-state measurement for such complex situations. So there is a lot of work to be done for the interested scientist.

But there are good reasons to be optimistic. In the last few years, experiments have made immense progress, and a lot has been achieved that was unimaginable to early generations of physicists. One of the reasons for hope is the simple fact that it was possible to observe quantum interference even for large molecules consisting of hundreds of atoms, like buckyballs (fullerenes) and their compounds. Buckyballs (named for Buckminster Fuller's geodesic domes) are carbon molecules that look similar to a soccer ball. Quantum interference has been observed for such objects with very high precision. Observation of quantum interference is, as we have seen, confirmation of quantum superposition, and thus it is an important first step toward entanglement. Today, nobody knows how an entangled state of two such complicated molecules could be made and how, even more challenging, two such molecules could be projected into an entangled state, as would be necessary in a generalized Bell-state measurement. But there is no reason why this should not be achieved in experiment someday, maybe even sooner than many people expect today.

A far-out experimental challenge is to show quantum phenomena for living systems. Schrödinger, for example, invented a *gedanken* (thought experiment) where he suggests creating a superposition of a live and a dead cat. There are many good reasons to believe that this is just science fiction in its best sense, the fiction of a new science. But on the other hand, there is no reason in principle why it should not be possible to observe quantum superpositions of living systems someday. For example, there is no fundamental reason why one should not be able to observe a quantum double-slit experiment for an amoeba or a very small bacterium. Surely, many experimental challenges would have to be surpassed first. For example, we would have to invent ways to shield such a small living creature from the hostile environment of the experiment. All of these experiments would have to happen in a vacuum. More generally, a system showing quantum interference should not interact with the general environment, as this would destroy its quantum state. Now, a living system does not like the idea of living in a vacuum or of being completely isolated. It needs nutrition, it needs oxygen from the air, it needs to live at a certain temperature, and so on. But if we just play the game of science fiction for a moment, we can easily imagine that using nanotechnology, we might build a small housing around the small bacterium or amoeba that protects it completely from the environment. Such a living being could then survive the flight through a vacuum apparatus and similar challenges. Again, this is science fiction, but there is no reason why we should not succeed someday.

TELEPORTATION AS A MEANS OF TRAVEL?

Science-fiction writers invented teleportation in order to be able to move people with the snap of a finger from one place to another. Obviously, this invention was very useful for covering the huge distances of space. From a practical point of view, it was even more important that in science-fiction films, filmmakers were able to save money using teleportation. Teleportation can be done simply and cheaply in the cutting room. Thus film companies can avoid the expense of simulating a spaceship landing on the surface of a strange planet—and the special effects needed to show that—which would certainly be a significant part of the expenses of making a film.

The idea of teleportation is attractive to people and therefore was in itself a giant success. But what does the "reality" look like? Did the experiments that we discussed in this book bring us closer to the possibility of teleporting people?

The answer is an emphatic *no*.

To see this, let us make a short list of what we would need to teleport people. As you will see, that list is a list of impossibilities.

1. The person to be teleported must be in a quantum state. A quantum state can only be achieved when the system to be teleported is completely isolated from the environment. Most interactions with the environment would destroy the quantum state of our human system and thus prevent us from being teleported. The essential point here is that such a quantum state for a living person would have to include superpositions of different states, but it is not even clear what that could mean. Schrödinger illustrates this in the case of a cat that is in a superposition of the two states "dead" and "alive." Clearly, no one knows what the meaning of such states would be, and furthermore, no one has any idea how such states could be produced or how to measure them in experiments. Another problem is that a person to be teleported has a mind, a consciousness, and maybe a soul. Using that mind, the person to be teleported would watch his own environment. That observation alone might be enough to destroy the quantum-mechanical superposition, since it could provide the person to be teleported with information about which of the different superposed states the person actually is in. In conclusion, all this would make teleportation impossible.

2. But let us assume that we can overcome all the problems mentioned under point one, and thus let us assume that it has become possible for us to put a person into a quantum state.

According to the quantum-teleportation protocol, the next challenge now arises. We have to produce an additional entangled twin pair of persons. Remember that entanglement does not just mean that the two members of the entangled pair are like identical twins. Entanglement means that neither of the two entangled systems carries any properties on its own before being measured. Before being observed, neither of the entangled twins carries its own hair color or eye color, or any other individual properties. Certainly, all other individual features must also be undefined. Yet, the two entangled persons must be perfectly correlated in the sense that when we observe one and see that the person assumes a certain hair color, a certain eye color, and so on, the other one, its quantum twin, would immediately be projected into a state with the same properties, no matter how far apart the persons are. This must hold for all features of the teleported persons. Not only does this sound strange, there is no way to make sense out of it in terms of constructing some experimental protocol concerning how to realize such entangled states. Needless to say, such entanglement of persons would also raise enormous ethical problems. Who on Earth would want two people entangled with each other in such a way that they do not have any well-defined properties? And who would want their existences as separate individuals to depend on the mercy of some future experimenter who decides or does not decide to perform measurements on them? And who would want to accept the unavoidable certainty that it is completely random which properties these entangled persons would achieve upon measurement? Obviously, we are talking a lot of nonsense for any person with a reasonable mind.

3. But suppose we succeed in producing entangled persons. The worst is still to come in terms of the experimental challenge. We finally have to project the person to be teleported and one of the two entangled ones into an entangled state. This is the generalization of the concept of a Bell-state measurement as we discussed it in our teleportation experiments with photons. No one has the foggiest idea how this would

work. Mathematically, we can in principle write down what such states might look like, but that is just playing mathematical games. These questions are so far out of the reach of the imagination that one cannot even talk ironically about the issue, as we did it in points one and two above.

In conclusion, it would not be wise for us to hold our breath or delay a faraway trip in order to wait for teleportation to work for people. Travel by teleportation remains a nice idea for science fiction.

That quantum teleportation most likely will be used someday to transfer information between quantum computers as we discussed above is a very different question.

SIGNALS OUT OF THE SKY ABOVE TENERIFE

Our small car climbs up the winding road through scattered woods. It is May, and colorful flowers enrich the green of the woods and the reddish brown of the rocks around us. It was only half an hour ago that we left the airport on Tenerife in the Canary Islands, and we have already reached an altitude of about 2,000 meters (6,500 feet). Suddenly, the landscape changes completely. We arrive on a high plateau with bizarre rock formations. In front of our eyes rises El Teide.

"This is the highest mountain in Spain," explains Zoran. And our young Spanish friend Josep adds, "Every Spaniard is proud when he has climbed the 3,718-meter-high peak of El Teide."

"Even though it is only a small distance from the top station of the cable car to the summit," Zoran adds, smiling, and thus dampens my admiration for the unknown mountain climbers significantly.

Zoran Sodnik and Josep Perdigues work as scientists for the European Space Agency (ESA). They are responsible for the scientific and technical program in a newly developing field: optical communication with satellites.

We pass an interesting rock formation called La Catedral. It tells us a story of many thousands of years ago, when El Teide was an active volcano that created the island of Tenerife out of the floor of the sea.

"But you can't just go up with the cable car and climb to the summit like that," Josep explains. "You need a license to go to the peak, and that license can only be obtained in a certain office down in Santa Cruz on the sea. So you have to work to get that license." My admiration for the unknown mountain climbers rises a little, but maybe in the wrong way.

Zoran smiles. "But you also can circumvent that license. The guy who checks the licenses goes up in the morning at nine o'clock with the first cable car. So if you start early enough and walk on foot all the way up instead of taking the cable car, you can pass the checkpoint before nine o'clock and thus go up to the peak without any problem. That's the reward for the really determined ones."

From down here, at the bottom station of the cable car, the distance from the top station at 3,550 meters up to the summit does not look very impressive, but climbing at an altitude of above 3,500 meters can be exhausting.

Our car keeps going on a curved road through the volcanic formations. Suddenly, after we take a particular curve, the scenery looks completely futuristic. We see at some distance half a dozen shining white, strangely formed buildings, some of which obviously are housing telescopes. We are at our destination: these are the telescopes of the El Teide Observatory, which is part of the Instituto de Astrofísica de Canarias (IAC), the Astrophysical Institute of the Canary Islands.

Most of the telescopes here were especially built to investigate various aspects of the Sun. They are being operated by a number of European countries that have been teaming up to create this conglomerate of telescopes. The institute has another branch on another Canary Island called La Palma, about 150 kilometers, or 93 miles, from Tenerife. The telescopes on La Palma are mainly used for research on the night sky. Because of the high population density on Tenerife, the sky at night is not dark enough for scientists to see faint objects in the sky. Therefore, many of the telescopes here are used for observation of the Sun.

"The first connection with Artemis is scheduled for half past nine in the evening, so we still have time to get something to eat," Josep proposes. "And it's about time, since I had only a small sandwich for lunch."

Now I cannot avoid the suspicion that Josep's way of driving, which was rather fast and energetic, and the fact that he drove faster and faster the closer we came to the institute, had something to do with the necessities of his stomach.

We move into our rooms inside the IAC, which reminds us of a typical chalet in the Alps or other mountain ranges: strong wooden beams and simple furniture, but a cozy atmosphere. The only differences are the satellite pictures on the walls in the observatory area and the telescopes outside the windows. We enjoy our simple meal, which we all—not just Josep—like very much. Afterward, we are ready to walk over to our real goal, the Optical Ground Station (OGS), operated by the European Space Agency. Zoran is responsible for OGS, and he is keen on realizing new scientific ideas there.

It is pitch dark when we step out of our lodgings. There are no street lamps, and no light whatsoever makes it out of the building we just left. Everyone up here tries to avoid producing any light, as this might disturb the operation of the telescopes.

But after a few steps, we realize that the small flashlight each of us carries is not necessary. After our eyes have adapted themselves to the darkness, the little bit of light shed by a very narrow slice of Moon is plenty to show us the way.

Soon we arrive at the OGS telescope, where we are welcomed by Eduardo and Martín, who are responsible for operating the instruments today.

Zoran unpacks two bottles of champagne that he has brought with him. "It is our tradition to open a good bottle after each successful connection with Artemis," he says. "We have about ten minutes left until the connection is possible. Let's walk over to the control room."

There, we learn that the Artemis satellite was put into space by the European Space Agency in order to test methods of optical communication between ground stations and satellites and from one satellite to another.

Usually, radio signals are used to communicate with satellites. These signals are on the one hand necessary for sending commands up to the satellite; on the other hand, the satellite sends information collected by its instruments down to Earth. Particularly when a satellite has to send down pictures, the amount of data is very large. In that case, it would be useful to have a means of communication that could carry more information than a radio signal.

In modern telecommunication on Earth, large amounts of data are indeed already sent by light. In modern cities, there are glass-fiber cables everywhere, connecting different computers. Such an optical connection will also be useful when a satellite has to send pictures down to Earth. That is exactly the purpose of Artemis—to check how that can be done and to see what is needed to do it successfully.

"At exactly nine thirty, Artemis should switch on its beacon of light. This is a small laser, not much stronger than a flashlight, that will shine down to us on Earth from a distance of 35,000 kilometers, or 22,000 miles. This is as far away as Artemis is," Zoran explains.

"But how does Artemis know whether its light hits us?" I ask.

"That's exactly the point," Zoran says. "Artemis knows roughly where we are, and then it lets its laser beam cross the area where we might be in a zigzag way. Using our telescope, we look to Artemis, and as soon as we happen to catch the light, we tell Artemis to stop its zigzag motion. Then we know that the beacon light has arrived at our place. In the next step, Artemis switches over to a much smaller light beam, which is then used to transmit the data."

It is now 9:30, but nothing happens. Shortly after, we see on one of the computer screens a small moving spot of light. The laser beacon of Artemis is shining down on us. The computers send up the command to stop the search motion, and the spot of light remains fixed. We quickly step out of the control room into the cupola housing the huge telescope. It is looking up into the sky at a very steep angle right now.

"Should we be able to see the light coming from Artemis?" I ask.

"Not quite," Zoran responds. "For our eyes, the light is not visible. It is infrared, a color our eyes are not able to see." He passes me a night vision device, which looks just like binoculars. It works like a digital camera, as it catches the infrared light and transforms it into an image on the screen that the human eye is able to see. And suddenly, here is the dot that comes from Artemis. This is light, made by humans, that reaches us from a distance of more than 35,000 kilometers!

Later on, we open up our bottle of champagne and celebrate our successful connection to the satellite. This is exactly the technology with which we want to perform our future experiments on entanglement and quantum teleportation. We are hoping to someday launch a satellite that will be a quantum successor to Artemis, a next-generation satellite that will contain a source of entangled photons.

The big challenge will be to teleport the quantum state of a photon from a satellite to Earth or vice versa. The fundamental idea there is quite similar to one in the experiments discussed earlier in this book, but the experimental realization will be much harder, because the devices on board a satellite must be extremely reliable and should not fail under any circumstances. If something fails in the laboratory on Earth, we can go there and fix it. This is impossible on a satellite.

We spend the night taking further test measurements, which are all very positive. The next day when we leave for the airport, we are all quite happy. We have seen with our own eyes that a future experiment involving photon entanglement via satellites is feasible in principle. Many years of interesting scientific research lie in front of us.

The car leaves the volcanic plateau of El Teide, and after passing the scattered pine forests, we return to the modern world at the coast of this beautiful island.

RECENT DEVELOPMENTS AND
SOME OPEN QUESTIONS

While this book is being finished, numerous experiments on quantum computation, quantum teleportation, and similar topics are going on in many laboratories all over the world. I am sure that in the time between the last lines of the book being written and the moment when you, the reader, have this book in your hands, a lot of new developments will have happened.

One of the most interesting developments is the preparation for space experiments. Part of these preparations are experiments using the telescope of the Optical Ground Station (OGS) operated by the European Space Agency on the island of Tenerife in the Canary Islands. In the experiments that have been performed so far and that are still going on, one station is on the island of La Palma and the other one is on Tenerife. The two stations are separated by 144 kilometers, or 90 miles. On La Palma, we have a small station and a platform where entangled photons are created. One of the two photons is measured locally on La Palma; the other one is sent over to Tenerife. It is difficult to catch these individual photons over such a large distance. One of these experiments is a nice example of an international collaboration with teams from the University of Munich, the University of Bristol, the University of Padua, and my group in Vienna.

Part of the challenge of catching photons over such a large distance is that the atmosphere is not stable. You can see that when you look at night up at the stars or over the ocean, for example, at a ship far away. You can see the light twinkling and also moving around a little. In the case of our photon experiments, this means that a photon starting on La Palma does not always meet the receiving station on Tenerife. One reason for the success of our experiment was that we built in an active correction mechanism. An additional beacon laser, on the OGS, very much like the beacon laser out on Artemis, shines over to La Palma, and likewise, there is one in the reverse direction. The sending station on La Palma and the receiving telescope on Tenerife both are constantly redirected such that the signal is maximal. So far, it has been possible to actually show that the photons are still well entangled after that distance, and to perform an actual quantum cryptography experiment.

In a parallel experiment, we use a similar telescope in Matera near Bari in the south of Italy. This is a collaboration with a group from the University of Padua. In that experiment, we send a faint laser pulse up to Ajisai, a Japanese satellite. That satellite consists of many cat-eye mirrors reflecting the light back to Earth. The goal of the experiment is to detect individual photons arriving down on Earth again. In the end, we made the laser beam going out so weak that typically for each of the light pulses sent up, only one photon was arriving back down on Earth. By precise timing, we were able to identify such individual photons. This is done by knowing exactly, from the position of the satellite, at which instant in time a photon should arrive back on Earth. The final goal of all these experiments is to prepare for quantum communication using satellites. The idea is to place a specially designed source on a satellite or on the International Space Station and send either one or two photons down to Earth to establish quantum teleportation and quantum cryptography over large distances, as these photons could be sent to places far away from each other.

A very important question concerns the development of quantum computers. There are many groups working worldwide in that field. Some use individual atoms or ions as carriers of information; others are working on using the standard semiconductor silicon technology of existing computers, modifying it such that individual quantum bits might be encoded and processed there. One idea is to implant individual atoms into silicon or another semiconductor one by one and have them talk to each other, which would constitute a quantum processor. Other groups work with small superconducting elements, and so on and so on. Today, it is completely impossible to foresee the further development of that technology and, in particular, to predict which technology will finally find industrial application. One point that is important for future development is that many of the concepts developed and demonstrated using one kind of physical realization can easily be transferred to another physical approach, for example, from atoms to photons or from photons to ions, since the basic underlying concepts, such as superposition and entanglement, are the same. So it might well be that the future quantum computer technology might be a hybrid combination of some of these ideas, or it might even be that we have not discovered the best way yet. New ideas still continue to come up in the scientific community all the time.

One of these fascinating new ideas was that of a one-way quantum computer, which we already mentioned briefly. The really fascinating fact here is that it works according to a completely different principle from all other computers, quantum or not. In a standard quantum computer, one feeds the input qubits into the quantum computer. The algorithm is then realized as a specific quantum evolution of these qubits.

A one-way quantum computer works in a fundamentally different way. Here, we start from a complex entangled state with many qubits. That state is extremely rich. It is so rich that in essence, it contains the solutions of all problems we might ask the computer to solve. As long as the state contains enough qubits, it is universal. The calculation now operates in a very fascinating way. The algorithm, that is, the procedure for performing the calculation, here is actually a sequence of measurements on that quantum state. It starts with the instruction of measuring a certain qubit in a specific way. That measurement projects that qubit onto a well-defined state. Remember, entanglement means that none of the entangled qubits has any state of its own. But when measured, it will randomly assume some property. This measurement of a single qubit breaks the entanglement of the measured qubit with the others. The measurement will also change the state of all the others entangled with it. So, doing a measurement on one qubit, we can drive the rest of the qubits to some other, still entangled, state. Then the algorithm tells us which qubit to measure next, and next, and so on. When all these measurements are performed and each gives the right result, we are left in the end with a few qubits that contain the result of the calculation.

One important problem is that each measurement on a qubit has a random answer. Now, it turns out that it's for only one of the two results that the rest of the qubits are projected into the desired state necessary for the calculation to proceed. In the other case, we simply have to discard the state and start again. Now, that is not a very efficient way to proceed. Luckily, as discovered by Raussendorf and Briegel, one can correct these mistakes by making the kind of the future measurement depend on which result was obtained in the first round. That way, such a quantum computer can be made deterministic. In 2007 our group actually performed such an experiment with entangled photons, which required very fast electronics for detecting a photon and feeding the result forward such that measurements on the remaining photons are changed fast enough. We found, as an initially unintended consequence, that it is possible to achieve that way a quantum computer that is faster than any other existing quantum computer concept.

From a conceptual point of view, a one-way quantum computer is quite interesting. In a sense, it is a quantum realization of the fictional Library in the short story "The Library of Babel" by the Argentinean writer Jorge Luis Borges. The Library consists of all books that have ever been written and all books that ever will be written. How could such a thing be possible? The idea is simple. The Library suggested earlier by Raimundus Lullus contains books with all possible combinations of letters. So, for example, it contains the book you hold in your hand now, but it also contains all possible books with one printing error, all possibilities with two printing errors, and so on and so on. In the end, you can imagine that such a library is quite useless. To find the right book is an incredibly complicated and essentially useless undertaking. You have to know its contents in every detail in order to find it.

In a sense, the quantum state with which the one-way quantum computer starts is something like this Library of Babel. This one quantum state contains all possible results of any calculation. Or in a sense, it contains all possible books. That in itself shows how rich quantum physics is. We now do not search for the right book, but by successive measurements on their states, we force the remaining qubits to be driven toward the desired result. This is a completely new idea of how computation works, and it might well fundamentally change our ideas about what computation means.

A fascinating idea obviously is a future quantum Internet, a worldwide network of quantum computers that exchange information using quantum teleportation. Such a quantum Internet could easily be made safe against any eavesdropping by means of quantum cryptography.

Whether or not such quantum computers will someday replace all existing computers is an open question. But there are reasons to be optimistic. And there is no fundamental reason why that should not be possible someday.

The question that is already leading to intense debates is whether or not quantum concepts could play some significant role in our own computer built into our head, namely our brain. There is general agreement on the role of quantum physics in all biological phenomena, namely, that the chemical processes happening are in the end quantum processes. But beyond that, there is no hint whatsoever that our brain uses, for example, qubits or even entanglement. The general opinion is that this would not be possible anyway, because the conditions in our brain are very different from those necessary to observe quantum phenomena. We remember that in order to observe entanglement and superposition, the system must be well isolated from the environment, since most disturbances from the outside world destroy the quantum states. That was, for example, the case in the double-slit experiment. Each disturbance of the particle, which would in principle allow us to find out which of the two slits the particle takes, destroys the quantum interferences. This phenomenon is called *decoherence*. A similar problem exists in entanglement. Disturbance of one of the two particles can very easily destroy the entanglement. The environment in our brain is very different. The nerve cells in the brain are immersed in a "warm soup," so to speak, and they are not at all isolated from the environment.

But from a fundamental point of view, we cannot exclude in principle the possibility that quantum physics might play some role in our brain. One hint why this question is not settled yet comes from the development of quantum computation itself. Very much to the surprise of many people, it has been discovered that even in the quantum computer, two possible mechanisms can be implemented to work against such disturbance. One is that it is possible to store information in a way that is robust to decoherence. This can be done such that the information is stored in properties or in degrees of freedom of the individual quantum systems that do not couple significantly with the outside. This is the approach of decoherence-free subspaces. Another way is to store the information jointly in many qubits such that it is in a sense redundant, and by doing a quantum comparison of these qubits, we can find out whether individual ones were changed as a result of some external disturbance and correct for that. This is the approach of quantum error correction. So it is not a priori inconceivable that similar mechanisms might also play a role in our brains, yet it is not clear today how and where in the brain such mechanisms might operate. So all this is just speculation today. But on the other hand, it would be a challenging research program to try to find out whether randomness, entanglement, or superposition plays any role in our brain. Some people believe that such questions could bring us closer to finding out what consciousness is, what the human mind is. Whether this is true or not is open. The questions "What is consciousness?" "What is the human mind?" "Will there ever be machines having consciousness?" "How can we find out whether a system, a machine, a living being has consciousness?" are questions that cannot be answered today in any definitive way. They will certainly be topics of intensive research in the future.

WHAT DOES IT ALL MEAN?

More important than these questions are probably the conceptual and philosophical consequences of quantum physics. We have seen in this book that some of the ways we are inclined to view the world simply do not work. We have learned that the idea that the world exists in all its properties independent of us, independent of the kinds of observations we perform, is in trouble. We are not just passive observers. The Austrian physicist Wolfgang Pauli expressed this by saying that the picture of an observer who is detached from the world does not work anymore. A detached observer would be just like a person in a theater watching a play taking place on stage. The question of whether he watches the stage or looks down on the floor does not at all change what happens on stage.

We have learned in this book that the observer has a significant influence through his choice of the measurement instruments, through his decision of what to measure. The point is that his measurement instruments don't just influence or change the observed systems. That would still be acceptable in some way. But we have learned that the choice of measurement instrument actually defines the property of a quantum system that becomes realized as an experimental result. For example, whether the observer doing a double-slit experiment chooses a setup that allows him to find out the path taken, or a setup that allows him to obtain the interference pattern, then decides whether path or interference pattern will be an element of reality. Yet a note of warning and caution is necessary here. It is dangerous—and not supported by the physics of the quantum measurement process—to claim, as is sometimes claimed, that it is the mind of the observer that influences the quantum state.

We also learned that a specific philosophical view has been definitely ruled out by the experiment. This is the concept of local realism. Local realism is the point of view that whatever we observe is defined in some way by a real physical property of the observed systems, a property that exists before and independently of our observation. Furthermore, local realism assumes that there are no instant actions at a distance. It assumes that what we observe is independent of what at the same time someone else far away decides to do, which measurement he performs on a distant particle entangled with our own, or whether he decides not to perform such a measurement of all.

We also learned that the quantum world is governed by a qualitatively new kind of randomness. The individual measurement result is purely random, without any possibility of detailed causal explanation. It is not just that we do not know what the cause is. This is probably the most fascinating consequence in quantum physics. Just imagine: centuries of scientific research, centuries of the search for causes, and attempts to explain why things happen just the way they happen lead us to a final wall. Suddenly, there is something, namely the individual quantum event, that we can no longer explain in detail. We can only make statistical predictions. The world as it is right now in this very moment does not determine uniquely the world in a few years, in a few minutes, or even in the next second. The world is open. We can give only probabilities for individual events to happen. And it is not just our ignorance. Many people believe that this kind of randomness is limited to the microscopic world, but that is not true, as the measurement result itself can have macroscopic consequences.

While we have learned here that local realism is untenable, the question is whether it is realism or locality that is not correct. So in other words, do we have to give up locality or do we have to give up the concept of realism? Do we have to allow Einstein's instantaneous "spooky action at a distance," and can we in that way save realism, or do we have to give up a realistic picture of the world even if we are willing to give up locality? These kinds of questions sounded open until recently, when Tony Leggett of the University of Illinois at Urbana-Champaign proposed something very interesting. He suggested a model according to which nonlocality is allowed. That means that instant action at a distance is permissible as long as it does not allow signaling faster than the speed of light. And then, he showed that a whole class of reasonably realistic theories are actually in conflict with quantum mechanics. In a very recent experiment, my group, in collaboration with Marek Zukowski, actually demonstrated that the predictions made on that model are in conflict with experiment. The conclusion therefore is that accepting nonlocality would only save realism at a high price, namely that the world that we consider realistic has very strange properties. Further discussion of these points would certainly go beyond the scope of this book, and it happens that the philosophical ramifications are not at all understood at present.

So in general, we have to conclude that while some commonsense pictures of the world are not tenable anymore in the view of quantum physics, it is not really clear how a new view of the world would work. One point is clear. The predictions of quantum mechanics are so precisely confirmed in all experiments that it is very unlikely, to say the least, that quantum mechanics is an incorrect description of nature. So we might now speculate a little bit about what such a new view of the world might look like.

A new picture of the world must encompass three properties that evidently seem to play a significant role in quantum experiments. The first two have to do with freedom. We might interpret the objective randomness of the individual quantum event as a freedom of nature. Nature gives us the answer it likes freely, without any predetermined cause. The fact is that the individual measurement with few exceptions is not determined in any possible way, not even in a hidden way. The second important property of the world that we always implicitly assume is the freedom of the individual experimentalist. This is the assumption of free will. It is a free decision what measurement one wants to perform. In the experiment on the entangled pair of photons, Alice and Bob are free to choose the position of the switch that determines which measurement is performed on their respective particles. It was a basic assumption in our discussion that that choice is not determined from the outside. This fundamental assumption is essential to doing science. If this were not true, then, I suggest, it would make no sense at all to ask nature questions in an experiment, since then nature could determine what our questions are, and that could guide our questions such that we arrive at a false picture of nature. Again a note of caution is necessary here. It does not at all follow that quantum randomness explains free will, as is often stated.

So I suggest that these two elements of freedom must be essential elements of our future picture of the world. But there is a third element, which is at least as important. That is the notion of information. Information has a significant role in quantum physics, and that role seems to go beyond the role it plays in classical physics.

Again, in classical physics, as suggested by Pauli, we have the picture of the detached observer. In that picture, the information we gain about a situation is derived from the world, from its properties; it is secondary; it is information about something that already exists, even if the process of gaining the information might change the properties of the observed system. It is the information a detached observer acquires.

The situation in quantum physics is quite different. We learned, for example in the double-slit experiment, that a decisive criterion for interference is whether any kind of information leaks out about the path the particle took, whether it passed through one slit or the other one. If that information is somewhere—if it's possible at least in principle to obtain knowledge of which path the particle took—no interference pattern arises. It is when there is no information present, not even in principle, about the path taken, independent of whether we take notice of it or not, that the interference pattern, and thus quantum superposition, occurs. So, information plays an interesting dual role. It is what we can know in principle if our means of extracting the information are good enough, if our technology is advanced enough, and so on. But it is also the possibility of making a statement about the world that determines what can be the case.

So it seems that what we can say in principle about the world has a crucial influence on the elements of reality. It was just the same in our teleportation experiment. There also an essential notion was that of information. The quantum state that is being teleported is nothing other than information.

Thus not only does what we can say about the world play a significant role in forming our picture of the world, but it also plays a much deeper role in defining what can be an element of reality, in the sense of which features can manifest themselves in an experiment as reality.

We can now make a very important observation. This is the observation that the concepts *reality* and *information* cannot be separated from each other. It is not possible even to think about reality without using what we know about reality, that is, information. In the history of physics, we have seen that significant progress happened when we gave up on separating concepts that people had thought of as being completely distinct. An important advance was, for example, giving up on the separation of the concepts *space* and *time* in relativity theory and unifying them in one joint notion called space-time. It seems that two other notions of that kind are the notions of *information* and *reality*. But what a future picture of the world where these two notions are something like two sides of the same coin looks like is very much open.

It now becomes clear why Einstein had to criticize quantum mechanics, why he called entanglement "spooky." His picture of the real, factual reality that exists in its essential properties independent of us, this picture of a separation of reality and information, does not seem to be tenable in quantum physics.

So, in conclusion, our world on the one hand is freer than what classical physics would allow us. On the other hand, we are also embedded in a stronger way into the world than what was the case then.

APPENDIX

GLOSSARY

INDEX

APPENDIX

ENTANGLEMENT—A QUANTUM PUZZLE FOR EVERYBODY

A. Quantinger

ABSTRACT

One of the most interesting phenomena in quantum physics is entanglement. Albert Einstein expressed his dislike of this phenomenon by calling it "spooky." Entanglement describes the phenomenon that two (or more) particles (or systems) may be so intimately connected to each other that the measurement of one instantly changes the quantum state of the other, no matter how far away it may be. These connections cannot be explained by properties these particles carry locally for themselves. John Bell showed that predictions of such local realistic theories are in conflict with those of quantum mechanics. It is the purpose of this paper to present a discussion of entanglement accessible to the general public. This is done by presenting both the typical experimental situation and the arguments leading to Bell's inequality. The paper concludes with a brief discussion of possible philosophical consequences.

INTRODUCTION

Quantum physics was created in the first quarter of the twentieth century in order to describe the behavior of atoms and other microscopic particles, including specifically photons, particles of light. Quantum physics today is an extremely important and highly successful description of nature. Its applications include, for example, the transistor, and therefore all modern computer chips, and the laser, to name some technological examples. It addresses elementary particles as well as the physics of the early universe. Also, the quantum mechanical description of nature is mathematically beautiful and exact. All its mathematical predictions have been verified to utmost precision in experiments.

Yet, while quantum mechanics is a very successful theory, there is still a problem. This problem is of a conceptual nature. Some of the predictions of quantum physics question central cherished aspects of our view of the world. In the general public, notions like Heisenberg's uncertainty principle and "quantum leaps" are well known. But the most interesting phenomenon is *entanglement*. The name entanglement was created by the Austrian physicist Erwin Schrödinger, who called it the *essential* notion of quantum mechanics, the one that forces us to say farewell to all our cherished views of how the world works. We will now discuss entanglement in more detail.

In 1935 Albert Einstein, together with Boris Podolsky and Nathan Rosen, published a paper[1] with the title "Can Quantum-Mechanical Description of Physical Reality Be Considered Complete?" In that paper, known as the EPR paper, the scientists show that according to quantum physics, two systems can be connected in an extremely close way, much closer than what is possible for systems in classical physics.

Let us consider two particles that had some interaction with each other. They might, for example, have collided at some earlier time. After the collision, the two particles fly away from each other. The EPR paper shows that quantum mechanics predicts that measurement on one of the two particles changes the quantum state of the other one, independently of how widely apart the two particles are separated. This influence exerted by the measurement of one particle on the other one happens instantly, without time delay. This seems to be in conflict with Einstein's own theory of relativity, according to which nothing can be faster than the speed of light. Einstein called this influence phenomenon "spooky action at a distance." He hoped that it might be possible to invent a new physics where such spooky actions do not happen.

Immediately after the publication of the EPR paper, Erwin Schrödinger also considered[2] the phenomenon, and he invented the name "entanglement."

The EPR paper was essentially ignored by most physicists for a long time. People were happy that quantum mechanics gave such an exact description of nature, and they were busy applying it to all kinds of phenomena. The situation changed radically in 1964, when the Irish physicist John Bell published a paper[3] with the title "On the Einstein-Podolsky-Rosen Paradox." In that paper, Bell showed that it is not possible to understand the phenomenon of entangled systems if one starts from rather "reasonable" assumptions of how the world should work, assumptions that one might even be tempted to call self-evident.

Bell's theorem might be one of the most profound discoveries of science since Copernicus, as the American physicist Henry Stapp once remarked. Copernicus changed the old picture of the world, according to which Earth was the center of the universe. Bell delivered a death blow to the local realistic picture of the world. Yet, there is an important difference. Copernicus showed us at the same time a new picture of the world, where the planets circle around the Sun. In the case of quantum physics, that new picture of the world is still in the making.

In the time since Bell's work in 1964, many experiments have demonstrated that the predictions of quantum mechanics for entangled particles are fully correct. So these experiments confirm that the world is really as "crazy" (Daniel Greenberger) as predicted by quantum mechanics. While Bell's considerations and the related experiments were essentially motivated by scientific curiosity, something happened that surprised everyone who was participating in the early experiments. These experiments unexpectedly laid the groundwork for ideas about a new information technology. The most important concepts in quantum information technology are the quantum computer, quantum cryptography, quantum communication, and quantum teleportation. Many believe that these are the cornerstones of the information technologies of the future.

PERFECT CORRELATIONS AND EINSTEIN, PODOLSKY, AND ROSEN

In science in general, and in physics especially, we like to describe nature quantitatively. The process often is that we make an observation and then try to understand what the reason for the observed phenomena might be. The general goal is to find a complete theoretical description in the language of mathematics. The mark of a successful theory in physics is for it to be able to make predictions for future observations. These predictions can then be scrutinized in experiments. In general, a theory is considered valid as long as it has not been contradicted by experiment.

Let us consider (Figure I) a source S that emits pairs of particles.[4] One particle—we can call it particle a—flies to measurement station A, the other particle, particle b, to measurement station B.

The measurement apparatuses at both stations A and B are completely identical. Each one contains some inner mechanism, which need not interest us here. It suffices us to know that with each apparatus, we can perform three different kinds of measurements on the incoming particle. Which of these measurements is performed is determined by the experimentalists, who each operate their own measurement station. The experimentalists are able to decide which of the three measurements are performed. This is done by placing the switch on the apparatus into one of three possible positions, x, y, or z. A further important feature of the measurement apparatus is that on each side, only two possible measurement results can occur. We call these measurement results + and −. Furthermore, we assume that each particle emitted by the source S is indeed registered in its respective apparatus. That means that both particle a and particle b will deliver either the result + or the result −, whatever the chosen position of the x, y, or z may be. The source emits one pair of particles after the other, but never two pairs at the same time.

An important experimental observation is that for any given setting x, y, or z on either side, A or B, the measurement results + and − occur with equal probability. This means that if we measure many particles, both on side A and on side B the result + will occur in about half of the cases and the result − in the other half. For many particles registered, the sequence is apparently random. A typical sequence of measurement results could be

$$+ - - + - + + - \ldots$$

Measurements on one side alone, on A or B, are called single-particle measurements. Our first conclusion from the experiment is that such single-particle measurement results have no structure whatsoever.

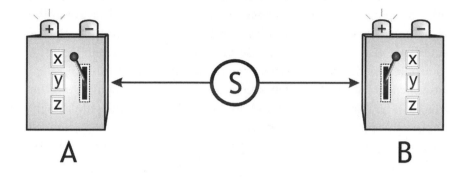

Figure I. Arrangement for the experimental observation of entanglement. A source S emits pairs of particles. One particle is registered by measurement station A, the other particle by measurement station B. Using a switch at each measurement station, the experimentalist can decide which kind of measurement, x, y, or z, is performed on the respective particles. The measurement result for each of the positions of the switches has only two possibilities, + or −.

Since the particles are created in pairs, it is reasonable to study which result, + or −, on measurement station A appears together with which result on measurement station B. That is, we study the correlations of the measurement results on both sides. Which measurement result at A is correlated with which measurement result at B? This is quite simple. We just have to check which measurements occur at the same time, since the two particles *a* and *b* of any given pair are created simultaneously and the two measurement stations are separated from the source by the same distance. Such simultaneous events are called *coincidences*. A result of such a coincidence measurement could, for example, be that at measurement apparatus A, where the switch might happen to be on position *x*, the result + occurs and at the same time, at apparatus B, where the switch might happen to be on position *y*, the result − occurs.

The question we have to ask now is, Which result + or − on one side occurs together with which result + or − on the other side, and how does this depend on the choice of switch positions?

As a first step, we restrict our considerations to situations where on both measurement stations, A and B, the same settings of the apparatus are chosen. There are three possible choices on each side, *x*, *y*, and *z*. So the respective pairs of identical settings for both sides are *x-x*, *y-y*, and *z-z*. For those specific situations, that is, when both settings are the same, experimental observations have shown that for each individual pair, we get the same result on both sides, at apparatus A and at apparatus B. So, if both apparatuses are set to perform the same measurement, we get the result + + or − −. Different results, + − or − +, never occur. This is true for any of the three combinations of settings, *x-x*, *y-y*, and *z-z*. Furthermore, both possibilities, + + and − −, occur equally often, independent of the switch setting *x*, *y*, or *z*.

These observations lead to a very important conclusion. On the basis of a specific result obtained on one side, say, at apparatus B, it is possible to predict with certainty the result of the measurement performed by apparatus A if the switch positions are the same.

Einstein, Podolsky, and Rosen suggested that there must exist an element of reality that corresponds to the measurement result, whenever it is possible to predict that result with certainty. This is called the EPR *reality criterion*.

In principle, the perfect correlations could arise because of some unknown kind of communication between the two apparatuses A and B. Such a communication would, for example, mean that apparatus A, when it measures its particle a, sends a message to apparatus B telling it what its switch position is and which measurement result occurred. Then apparatus B would simply provide the same measurement result if its switch is at the same position. In order to rule out such an explanation, the EPR paper assumes that the two apparatuses are so widely separated that the information cannot arrive in time, as it cannot travel faster than the speed of light. It is a consequence of Einstein's own theory of relativity that no signal can be faster than the speed of light. So the EPR paper requires that the measurement result on one side cannot depend on what is done at the same time to the particle on the other side—which measurement is performed on it or even whether or not it is measured at all. This is in essence the *locality assumption* of Einstein, Podolsky, and Rosen.

A theory that obeys both the reality criterion and the locality assumption of Einstein, Podolsky, and Rosen is called a *local realistic theory*.

We will now see that the perfect correlations found above can be explained by a very simple model based on the ideas of local realism.

Our model assumes that each particle carries some property or instruction that determines the specific measurement result, one specific instruction for each possible measurement setting. The instructions carried by both particles must be identical for the same setting on both sides in order to explain the perfect correlations. But they may vary from one setting to another. In the spirit of the EPR criterion, the assumption of such properties or instructions is rather reasonable. These additional properties completely explain why one can predict with certainty what the result on the other side, say B, will be once one knows the result on A for the same setting. It is simply that both particles carry identical pairs of instructions.

We call these additional properties each particle carries *hidden variables*, as they need not be accessible to direct observation. For our model, it suffices us to assume that all they do is determine the measurement result on their respective sides.

So far, we seem to have a simple and rather successful model to explain the measurement results. But any good model also has to explain other situations it was not invented for. Or at least, it should not be in conflict with observation in other situations. In our case, these are situations where the switch positions at station A and station B are not the same—situations where we have to allow all possible combinations of x, y, and z between both sides. It does not follow from our model that the results on both sides must now be the same. Our model only predicts this feature if the two switch positions are identical. So now, the results cannot only be $+ +$ or $- -$, but also $+ -$ or $- +$. It is evident that our model, which was invented for perfect correlations, is not rich enough to predict exactly how often these two other possibilities occur.

Nevertheless, Bell was able to show that these combinations cannot occur arbitrarily often. There are limits for how often they occur if one assumes the model we discussed. These limits are given by *Bell's inequality*.

BELL'S INEQUALITY FOR NON-PHYSICISTS

To make Bell's inequality easily accessible for the general readership, we will translate its language into that of everyday experience. The argument follows in essence a paper by Eugene Wigner[5] that builds on Bell and is expanded by Bernard d'Espagnat.[6] Instead of pairs of particles, we look at identical human twins. Then the three measurements x, y, and z of the particles correspond to the observation of three features of the twins—say, their height, hair color, and eye color. The measurements performed on the twins are simply visual observations. We observe whether they are tall or short, whether their hair is blond or brunet, and whether their eyes are blue or brown. We disregard any twins who have different properties, for example, other hair or eye color. So our measurement results again have two values for each observation.

Our identical twins exhibit perfect correlations just like those we discussed for pairs of particles. For example, if one of the twins is tall, blue-eyed, and brunet, we know that the other twin will also be tall, blue-eyed, and brunet. According to Einstein, Podolsky, and Rosen, these three properties—height, eye color, and hair color—are elements of reality that we can predict with certainty for the second twin upon observation of the first twin. We also know the reason for these correlations: the twins carry the same genes. These genes correspond to the local hidden variables we just considered.

If we now look at a large number of such twin pairs, we get all possible combinations. For the three properties that we picked out, there are eight combinations:

- Tall, blue-eyed, brunet
- Tall, blue-eyed, blond
- Tall, brown-eyed, brunet
- Tall, brown-eyed, blond
- Short, blue-eyed, brunet
- Short, blue-eyed, blond
- Short, brown-eyed, brunet
- Short, brown-eyed, blond

So, of all the many pairs of twins that we are considering, some number will be tall, with blue eyes and blond hair, another number will be short, with brown eyes and brunet hair, and so on. How many we have in all eight possibilities, we do not know. But we do not need to know that. We are able to make some very simple statements. For example:

$$
\begin{pmatrix} \text{Number of tall} \\ \text{pairs of twins} \\ \text{with blue eyes} \end{pmatrix} = \begin{pmatrix} \text{Number of tall} \\ \text{pairs of twins} \\ \text{with blue eyes} \\ \text{and brunet hair} \end{pmatrix} + \begin{pmatrix} \text{Number of tall} \\ \text{pairs of twins} \\ \text{with blue eyes} \\ \text{and blond hair} \end{pmatrix}
$$

This equation is completely self-evident. A tall twin with blue eyes in our model must have either blond or brunet hair. There is no other possibility. From that equation, we can derive an inequality for pairs of twins:

$$
\begin{pmatrix} \text{Number of tall} \\ \text{pairs of twins} \\ \text{with blue eyes} \end{pmatrix} \leq \begin{pmatrix} \text{Number of} \\ \text{tall pairs of} \\ \text{twins with} \\ \text{brunet hair} \end{pmatrix} + \begin{pmatrix} \text{Number of} \\ \text{blond pairs} \\ \text{of twins with} \\ \text{blue eyes} \end{pmatrix}
$$

The symbol $\leq$ means that the left-hand side is smaller than or at most as large as the sum on the right-hand side. How did we get from the equality to the inequality? Very simple. In the first bracket on the right-hand side, we relax the condition of the eye color; it is clear that the number of tall pairs with blue eyes and brunet hair in our sample is either the same as or smaller than the number of tall pairs with brunet hair irrespective of eye color. Likewise in the second term on the right-hand side: we relax the condition of height, so the same reasoning applies.

Let's now assume that for some reason we are able to observe only one property on each twin. In this case, we write down the equation we just obtained in a modified way:

$$
\begin{pmatrix}
\text{Number of pairs} \\
\text{of twins where} \\
\text{one is tall and} \\
\text{the other has} \\
\text{blue eyes}
\end{pmatrix}
\leq
\begin{pmatrix}
\text{Number of pairs} \\
\text{of twins where} \\
\text{one is tall and} \\
\text{the other has} \\
\text{brunet hair}
\end{pmatrix}
+
\begin{pmatrix}
\text{Number of pairs} \\
\text{of twins where} \\
\text{one has blond} \\
\text{hair and the other} \\
\text{has blue eyes}
\end{pmatrix}
$$

This is *Bell's inequality for twins*. It is obvious that it must be true, as we just have seen.

Before we go further into the discussion of Bell's inequality, let's recapitulate what we did so far.

We looked at three different features of identical twins (height, hair color, and eye color) and restricted ourselves to just two variants of each of these features (tall-short, blond-brunet, and blue-brown). We did not consider any other twins or any other features. Then we considered in which combinations some of these respective features might occur, and we arrived at Bell's inequality.

As innocent as the Bell inequality statement we just found might look, that is how important it is for modern physics. It provides a qualitative criterion for why entangled quantum states are fundamentally different from anything in classical physics. It is clear that Bell's inequality is true for all pairs of objects with identical features. All we have to do now is to translate Bell's inequality into a specific situation. We also have to restrict ourselves to features with two possibilities. If we do that, then in daily life, Bell's inequality will always turn out to be correct for any twin objects with identical features.

Let us now translate Bell's inequality into our experiment, discussed above, on pairs of particles. There, too, we had three different features of the particle we observed depending on the measurement positions— x, y, or z. And we had two results, + or −, which are perfectly correlated with each other, when on both sides, A and B, the same property is measured for the respective particle. So this situation is just like the one for our identical twins. All we now have to do is to translate the language we used for the twins into the language of our particle model. We use the following correspondences:

- The size corresponds to the property x: tall is translated into the result $+$; short is translated into the result $-$.
- Eye color corresponds to the property y: blue eyes are translated into the result $+$; brown eyes are translated into $-$.
- Hair color corresponds to the property z: brunet is translated into $+$; blond is translated into $-$.

We can apply the same approach to our pairs of particles now, as with the twins before, because of the perfect correlations and the EPR reality criterion. That is, if we measure one property, on one particle, we know that the other particle will carry the same property, should we observe it.

With that translation, we can obtain *Bell's inequality for pairs of particles*:

$$\begin{pmatrix} \text{Number of} \\ +\, +\ \text{results with} \\ \text{apparatus A on } x \\ \text{and apparatus} \\ \text{B on } y \end{pmatrix} \leq \begin{pmatrix} \text{Number of} \\ +\, +\ \text{results with} \\ \text{apparatus A on } x \\ \text{and apparatus} \\ \text{B on } z \end{pmatrix} + \begin{pmatrix} \text{Number of} \\ +\, -\ \text{results with} \\ \text{apparatus A on } y \\ \text{and apparatus} \\ \text{B on } z \end{pmatrix}$$

Thus, we directly translated Bell's inequality for identical twins into Bell's inequality for identical particles in our experiment. The question is now how pairs of particles will behave in the real world. Many experiments have been done by many groups. Nearly all of them were performed with particles of light, photons. We will discuss that case now in detail.

ENTANGLED PHOTONS

We will explicitly consider photons entangled in polarization. The polarization of light is a property that is also known from everyday life. It describes the way light oscillates, horizontally (back and forth), vertically (up and down), or along some other direction. Photographers, for example, use polarization filters in order to take out reflections or glare in their pictures.

Individual particles of light, photons, also carry polarization. Let us take an individual photon and determine whether the photon is polarized along a certain direction or not. For the photon, there are only two possibilities. Whatever direction you pick, it will turn out to be either polarized parallel to this direction, which we call vertical polarization, or orthogonal to it, which we call horizontal polarization.

We now transfer our picture of particle pairs to pairs of polarized photons. It is rather easy in an experiment to create entangled pairs of photons where the polarizations of the two photons are closely connected, indeed entangled in the way Schrödinger meant it. There are different kinds of entanglement. The specific entanglement depends on the kind of source one uses. We will assume a simple case, namely, a source where the two photons always exhibit the same polarization if measured along the same direction. So both photons end up either horizontally or vertically polarized. Then the three measurements x, y, and z correspond to measurements of polarization along three different directions (Figure II).

If we use the combination of a polarizing beam splitter (PBS) and a rotatable half-wave plate (HWP), the polarization can be measured along any arbitrary direction. We will just consider three different positions of the half-wave plate. This means that we will perform measurements of polarization along three different directions. Let us call the results of the measurement of polarization along the first direction H and V, along the second direction H' and V', and along the third direction H'' and V''. So now, again, we have three different properties we can measure, the polarization according to the three different orientations of the polarizer. And we have two results, horizontal and vertical, for the polarization with respect to the chosen orientation.

We now can consider again two different cases. First, those cases where on both sides, A and B, the polarization is measured along the same direction, that is, where both half-wave plates (HWP) have the same orientation. Because of entanglement, we get the same result on both sides. So we will obtain one of the following six combinations: H-H, V-V, H'-H', V'-V', H''-H'', and V''-V''. This perfect correlation was our starting point for the derivation of Bell's inequality.

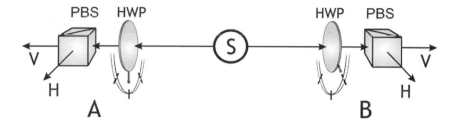

Figure II. An experiment for observing the polarization entanglement for pairs of photons. The source S creates pairs of photons. One of the photons is sent to measurement station A, the other photon to measurement station B. The polarization of each photon is being measured using a polarizing beam splitter (PBS). When the photon emerges in the H beam it is horizontally polarized, and likewise in the V beam it is vertically polarized. The measurement of polarization along different directions can be implemented using the half-wave plate (HWP). This half-wave plate can rotate the polarization by a certain angle, depending on its orientation. The measurement of polarization with a fixed PBS and a rotated HWP is the same as if one were to perform a measurement with a rotated PBS. That way, one can observe the polarization along any direction.

Let us now look at those cases where we choose different polarizer orientations on both sides. Then, we can translate the Bell's inequality that we have so far directly into new situations. We just have to translate the result H, H', or H'' into + and the result V, V', or V'' into −. The three orientations of the half-wave plate correspond to the three different settings of the switch, x, y, or z. Thus, we obtain *Bell's inequality for polarization entangled photons*:

$$
\begin{pmatrix}
\text{Number of pairs} \\
\text{where photon 1} \\
\text{exhibits} \\
\text{polarization } H \\
\text{and photon 2} \\
\text{exhibits} \\
\text{polarization } H'
\end{pmatrix}
\leq
\begin{pmatrix}
\text{Number of pairs} \\
\text{where photon 1} \\
\text{exhibits} \\
\text{polarization } H \\
\text{and photon 2} \\
\text{exhibits} \\
\text{polarization } H''
\end{pmatrix}
+
\begin{pmatrix}
\text{Number of pairs} \\
\text{where photon 1} \\
\text{exhibits} \\
\text{polarization } H' \\
\text{and photon 2} \\
\text{exhibits} \\
\text{polarization } V''
\end{pmatrix}
$$

Now, finally, we have achieved something very important. We have achieved an experimental prediction that can be directly checked. We are now left with two questions. First, do all predictions of quantum physics agree with the inequality that we just observed? Most interestingly, the answer is truly negative. There are sets of orientations for the polarizers, that is, in our case, orientations of the half-wave plates (HWP) where the above inequality is violated.[7] In these cases the right-hand side of the inequality is *smaller* than the left-hand side. So there is a contradiction between quantum mechanics and the arguments that led to Bell's inequality, that is, local realism.

The second question is, What happens in the experiment? Does nature agree with quantum physics, or does it obey the limitations implied by local realism? It turns out that photons in the experiment indeed do what quantum mechanics predicts. So far, there have been many experiments, and in all these experiments, with the exception of an early one, there is perfect agreement with the quantum mechanical prediction.

The assumptions that went into the derivation of the Bell inequality are the assumptions of local realism. So, the conclusion is that the philosophical position of local realism is untenable. A philosophical question about what the world looks like has thus been answered by experiment.[8]

WHAT COULD THAT MEAN?

How is it possible that a statement as simple as Bell's inequality might not hold in nature? The problem we have is that the considerations that led us to Bell's inequality were extremely simple. I would argue that they are so simple that the Greek philosopher Aristotle could already have derived Bell's inequality had he known that this was an interesting and nontrivial problem. We did not have to use quantum mechanics for its derivation. But Aristotle would never have expected that this could be an interesting problem. In contrast, he probably would have said that this is quite uninteresting, because nature obviously has to behave in a way so as not to violate the inequality. Let us just think again of our example with the identical twins, where we had a perfect explanation of the correlations between the twins.

Quantum particles do not behave like identical twins. Even if they always show the same results when they are measured for the same property, we are not allowed to explain this by saying that they carried that property before and independently of observation.

What kind of conclusions can we now draw from the violation of Bell's inequality? It is clear that at least one of the assumptions we used in its derivation must be wrong. What were these assumptions?

The first fundamental assumption was that of realism. This is the idea that an experimental result reflects in some way the features of the particles that we measure. The second fundamental assumption was the locality hypothesis. It is the assumption that the real physical situation of, say, measurement apparatus B including particle b must be independent of the kind of measurement done at the same time to the distant particle a using measurement apparatus A.

There is a third assumption, which we used implicitly but did not express in detail. It is the assumption that it makes sense to consider what kind of experimental result would have been obtained if one had measured a different property than the one that was actually measured. For the case of the twins, the assumption means that it makes sense to assume that, for example, blue-eyed blond twins must be either tall or short, even if we do not check their height. In the case of the measurement of two particles, it means that it makes sense to consider what the measurement result would have been, for example, for switch position z, even as the particle is measured with, say, switch position x.

We now discuss some of the possible conceptual consequences of the breakdown of local realism. One possibility is that the reality assumption is not correct. This would mean in principle that the property of a particle observed in a specific experiment is not an element of physical reality before the measurement is performed. In the end, this means that the reality depends on the decision of the observer—of the experimentalist—about which measurement to perform. The breakdown of realism would mean that the measured result does not reflect any kind of property that existed before and independently of observation.

Another possibility would be that the locality hypothesis is not correct. Such a breakdown of locality could, for example, mean that something is wrong with our picture of space and time. A quantum system that consists of two or more entangled particles remains an unseparated entity regardless of how far the individual components of the system are separated from each other.

A breakdown of the third assumption would mean that one is only allowed to talk about the properties of systems when these properties are indeed measured. Expressed very simply, the question "What if?" would be illegal. This would certainly contradict our everyday experience. We always consider different possible alternatives, and we base decisions on the possible consequences of these alternatives. For example, to know what will happen if we cross a superhighway during rush hour with our eyes closed, it is not really necessary for us to perform that experiment.

At present, there is no agreement in the scientific community as to what the philosophical consequences of the violation of Bell's inequality really are. And there is even less agreement about what position one has to assume now. Nearly all physicists agree that the experiments have shown that local realism is an untenable position. The viewpoint of most physicists is that the violation of Bell's inequality shows us that quantum mechanics is nonlocal. This nonlocality is exactly what Albert Einstein called "spooky"; it seems eerie that the act of measuring one particle could instantly influence the other one.

The other possibility would be for us to give up the picture of a world that exists in all its properties independent of us. That would mean that we have a very essential influence on reality just by deciding which measurement to perform.

There are indeed hints that this might be the message we have to accept. The most significant result in that connection is the so-called

Kochen-Specker paradox.[9] It would go too far to explain it in detail here. A brief mention of the result must suffice. The Kochen-Specker paradox can be stated rather easily. It says that even for individual quantum systems, if they are sufficiently complex, it is not possible to assign to them elements of reality that explain all possible experimental results independent of the full experimental context, i.e. which measurement is performed at the same time on the same system. Now, since Kochen and Specker only considered measurements on single quantum particles, the locality hypothesis does not come into play.

Just for completeness, let us mention that some other positions are also possible, at least in principle. One is the assumption of total determinism. In that case, everything is predetermined, including the decision of the observer about what he wants to measure. Thus, the question of what property the particle would carry if he were to measure something else would not come up at all, and therefore, the logical line of reasoning that led to Bell's inequality could not be carried out. It is obvious that such a position would completely pull the rug out from underneath science. What would it mean to do an experiment if that were the case? After all, an experiment is asking nature a question. If nature itself determines the question, then we might as well not ask that question at all.

Another logically possible position would be to assume that the individual measurements of the individual particles act back into the past. From that point of view, they would influence the source and tell the source, back in the past, with which properties to emit each particle. It is again obvious that such a position would mean a very radical rewriting of our views of space and time.

While we have to leave the answer to these philosophical questions open here, there are hints that they have to do with the role of information. Maybe it is true that the two concepts of information and reality cannot really be separated from each other.[10]

REFERENCES

1. A. Einstein, B. Podolsky, and N. Rosen, "Can Quantum-Mechanical Description of Physical Reality Be Considered Complete?" *Physical Review* 47 (May 15, 1935): 777.

2. E. Schrödinger, "Die gegenwärtige Situation der Quantenmechanik," *Naturwissenschaften* 23 (1935): 807, 823, 844. English translation: *Proceedings of the American Philosophical Society* 124, (1980): 323.
3. J. S. Bell, "On the Einstein-Podolsky-Rosen Paradox," *Physics* 1 (1964): 195–200.
4. N. D. Mermin, "Bringing Home the Atomic World: Quantum Mysteries for Anybody," *American Journal of Physics* 49 (1981): 940.
5. E. P. Wigner, "On Hidden Variables and Quantum Mechanical Probabilities," *American Journal of Physics* 38 (1970): 1005.
6. B. d'Espagnat, *Le réel voilé, analyse des concepts quantiques* (Paris: Fayard, 1994). English translation: *Veiled Reality, An Analysis of Present-Day Quantum Mechanical Concepts* (Reading, MA: Addison-Wesley, 1995).
7. For an overview see A. Zeilinger, G. Weihs, T. Jennewein, and M. Aspelmeyer, "Happy Centenary, Photon," *Nature* 433 (2005): 230.
8. For completeness, we note that in existing experiments, there are still some loopholes left. Yet, we assume that these loopholes will be closed in the near future.
9. S. Kochen and E. Specker, "The Problem of Hidden Variables in Quantum Mechanics," *Journal of Mathematics and Mechanics* 17 (1967): 59.
10. Hans Christian von Baeyer, "In the Beginning Was the Bit," *New Scientist* 2278 (2001): 26–33.

GLOSSARY

Bell's inequality   A mathematical expression derived by John Bell. It expresses the fact that correlations between two classical systems are limited in strength. Quantum mechanical measurements on entangled states are able to violate Bell's inequality.

Bell states   The concept that the polarizations of two photons can be entangled in four different ways with each other. These are the four maximally entangled Bell states.

Bell's theorem   The statement that entangled states and thus quantum physics contradict the view of local realism.

Classical physics   The realm of physics before quantum mechanics. There, objects may have well-defined properties and quantum uncertainty does not apply.

Double-slit experiment   An experiment where light, or any other particle, passes a diaphragm with two slit openings. The resulting particle distribution pattern on an observation screen depends on which kind of information exists about the path taken by the particles.

Electro-optical modulator   A device that rotates the polarization of light depending on the magnitude of an applied voltage.

Entanglement   The concept in quantum physics that two or more particles can be connected in a much stronger way with each other than in classical physics. Measurement on one can instantly, over an arbitrary distance, influence the quantum state of the other one. Albert Einstein called entanglement "spooky."

Entropy   A measure of the disorder of a physical system. It is given by the number of ways a specific situation can be arranged out of components. The more such possibilities, the higher the probability and the higher the entropy.

Heisenberg's uncertainty principle   The idea that quantum particles cannot be at a well-defined position and have a well-defined momentum (that is, speed) at the same time. If one is more certain, the other becomes more uncertain.

Heuristics   A method for finding physical laws or explanations by an intuitive approach based on common sense; a way of guessing a possible solution or explanation.

Hidden variable   The idea that quantum systems might carry additional properties not directly accessible to observation, but possibly explaining experimental results on a deeper level.

Interference fringes   The bright and dark stripes found on an observation plane behind a double-slit setup.

Laser   A light source of high intensity. In a laser beam, the light oscillates in sync.

Local realism   The assumption that the results of our observations correspond to a reality existing independent of our observation, and that there is no influence faster than the speed of light.

Malus's law   The law describing how the transmission of a polarized light beam through a polarizer varies with the angular orientation of the polarizer. Mathematically, this is the cosine law.

Particle   A particle is well localized in a single position and moves along a well-defined trajectory through space.

Photoelectric effect   The effect of light impinging on a metal plate, releasing electrons into space.

Photon   An elementary quantum particle of light.

Polarization of light   The way the electric field in a light wave oscillates is its polarization.

Probability   A measure of how frequent or how likely a specific experimental result is.

Quantum   Initially, each atomic or subatomic particle. Today, every system that shows quantum behavior such as superposition and entanglement. The plural is *quanta*.

Quantum complementarity   The feature that two or more observables of a quantum system—for example, the path taken by a particle in a double-slit experiment and the interference pattern—cannot be well-defined at the same time.

Quantum mechanics   As opposed to classical mechanics, the realm of physics that describes, originally, very small particles, but now, increasingly, larger objects. It is governed by notions like quantum uncertainty and entanglement.

Quantum superposition   The feature that a quantum system can be in two states at the same time, for example, two different spin states.

Quantum teleportation   The transfer of a quantum state—that is, certain properties of a system—over to another system, which may be in principle arbitrarily far away. Quantum teleportation uses entanglement as the means of transmitting that information.

Quark   An elementary constituent particle.

Random-number generator   A device that produces sequences of random numbers. These are used in various mathematical tasks.

Wave-particle dualism   The principle that photons—and other particles—can behave as waves or as particles depending on the experiment chosen.

INDEX

Page numbers in *italics* refer to illustrations.

World War II, 7, 132
  Nazis in, 99, 200